90 Topics in Current Chemistry

Fortschritte der Chemischen Forschung

W0246053

Plasma Chemistry II

Editors: S. Veprek and M. Venugopalan

Springer-Verlag
Berlin Heidelberg GmbH 1980

This series presents critical reviews of the present position and future trends in modern chemical research. It is addressed to all research and industrial chemists who wish to keep abreast of advances in their subject.

As a rule, contributions are specially commissioned. The editors and publishers will, however, always be pleased to receive suggestions and supplementary information. Papers are accepted for "Topics in Current Chemistry" in English.

ISBN 978-3-662-15974-3   ISBN 978-3-540-38980-4 (eBook)
DOI 10.1007/978-3-540-38980-4

Library of Congress Cataloging in Publication Data. Main entry under title: Plasma chemistry. (Topics in current chemistry ; 89–90) Bibliography: v. 1, p. ; v. 2, p. Includes indexes. 1. Plasma chemistry – Addresses, essays, lectures. I. Series. QD1.F58 vol. 89–90 [QD581] 540'.8s [541'042'4] 79–25770

© by Springer-Verlag Berlin Heidelberg 1980
Originally published by Springer-Verlag Berlin Heidelberg New York in 1980.
Softcover reprint of the hardcover 1st edition 1980

2152/3140–543210

Contents

Plasma Chemistry of Fossil Fuels

Mundiyath Venugopalan, Uptal K. Roychowdhury,
Katherine Chan, and Marion L. Pool

Department of Chemistry, Western Illinois University, Macomb, Illinois 61455, U.S.A.

Table of Contents

1 Introduction

Fossil fuels are the subterranean remains of green plants and animals that once grew and then were buried in sedimentary sands, muds and limes under conditions of incomplete oxidation. The present supply of fossil fuels includes coal, oil, natural gas, oil shale and tar sands. Natural gas is the fossil fuel in shortest supply and greatest demand. The simple hydrocarbon methane is the predominant component and represents 80–95 volume percent of any natural gas. Over the years several techniques have been applied to produce methane from other fossil fuels such as petroleum which is a mixture of hydrocarbons with six or more carbon atoms and coal which is a complex mixture of some organic compounds. One such technique is the production of a plasma in petroleum and coal through the action of either very high temperatures or strong electric fields. Since coal is in greatest supply the objective included obtaining new knowledge of coal chemistry, which may lead to new methods of producing organic chemicals.

The properties of the plasma produced in fossil fuels vary widely. Those plasmas labeled "glow discharges" are characterized by average electron energies of $1-10\,eV$, electron densities of $10^{15}-10^{18}\,m^{-3}$ and lack thermal equilibrium in the sense that electron temperatures are much greater than gas kinetic temperature ($T_e/T_g = 10-100$). The absence of thermal equilibrium makes it possible to obtain a plasma in which the gas temperature may have near ambient values while at the same time the electrons are sufficiently energetic to cause the rupture of molecular bonds. It is this characteristic which makes glow discharges well suited for the study of chemical reactions involving thermally sensitive materials such as petroleum and natural gas. By contrast, plasmas labeled "arcs" or "jets" have nearly identical electron and gas temperatures (> 5000 K). The high gas temperature makes these plasmas suitable for producing chemicals by degrading complex organic materials such as coal, shale and tar. The highly excited species that exist in these plasmas can react to produce compounds whose formation is thermodynamically unfavorable under ordinary experimental conditions. The physical and chemical properties and the production of both types of plasmas have been fully described elsewhere[1].

Because of the complex structure of coal and the variable composition of petroleum most of the plasma work using these materials is descriptive in nature. Attempts at modeling have been confined to the carbon-hydrogen system, chiefly using graphite, perhaps due to its importance in nuclear fusion and as aerospace material. In this chapter the studies of coal, petroleum hydrocarbons and natural gas in glow discharges, electrical arcs and jets are reviewed. Also reviewed are those studies in which these fossil fuel plasmas are formed in presence of simple gases such as hydrogen, nitrogen and argon. A comparison is then made with those studies in which lasers and flash heating techniques were applied. Pertinent investigations on the structural aspects of plasma-treated coal are included. Finally, the status of work on plasma desulfurization and gasification of coal and petroleum is discussed.

2 Thermodynamic and Kinetic Aspects of Fossil Fuel Chemistry

Thermodynamic considerations of the carbon-hydrogen system provide a useful guide to the nature and yield of products which might be obtained from fossil fuels at the temperatures attained in various plasma devices.

At temperatures between 900 and 2000 K most hydrocarbons have a positive free energy of formation, which, with the exception of acetylene, increases with increasing temperature[2]. Below 500 K only the paraffinic hydrocarbons are thermodynamically stable. Above 1700 K acetylene has a lower free energy of formation than the other hydrocarbons, but it is still thermodynamically unstable. Consequently, acetylene can be obtained by rapidly carbonizing fossil fuels at about 1800 K, but the yield is still mainly governed by chemical kinetics[3]. That is to say, the reaction time must be sufficiently long to permit the decomposition of other hydrocarbons to acetylene, but sufficiently short to prevent any appreciable decomposition of the acetylene formed to carbon and hydrogen. At temperatures of about 4000 K, the free energy of formation of acetylene from its elements approaches zero, and the equilibrium yield of acetylene is appreciable. The system is complicated, however, by other reactions and phase changes which occur at these high temperatures. For example, carbon sublimes at about 4000 K, molecular hydrogen dissociates, and various species such as C, C_2 and C_3 are formed.

Coals, particularly the bituminous and sub-bituminous varieties, undergo primary decomposition in the temperature range of 700–800 K. If coal carbonization could attain thermodynamic equilibrium over this temperature range, the hydrocarbon products with the exception of methane, if any, would be decomposed mainly to carbon and hydrogen. In practice, thermodynamic equilibrium is not attained, and the composition of the hydrocarbon by-products is mainly determined by the temperature and the kinetics of the process.

The equilibrium between carbon and hydrogen at high temperatures has been treated thermodynamically by several authors[4, 5]. The approach was to formulate the various reactions which could possibly occur, to apply to each the appropriate mass action equation, and to solve the set of simultaneous equations so obtained. A distinction was made between heterogeneous and homogeneous systems, since for the latter it is necessary to specify the mole ratio of carbon to hydrogen (C/H) in the system.

Assuming that the equilibrium composition of a reaction mixture with C/H = 1.0 at 2000–5000 K would consist of C, C_2, C_3, C_s, H, H_2, CH, C_2H, C_3H, C_4H, CH_2, C_2H_2 and C_4H_2, Baddour and Blanchet[5] found that the mole fraction of acetylene in the equilibrium mixture passes through a maximum with temperature, the value being 0.07 at 3300 K while that of C_2H is 0.1 at 3800 K. To apply this information to the products obtained at room temperature they assumed that C_2H_2 remained unchanged on quenching, while C_2H recombined with H to yield more C_2H_2. On this basis, the theoretical maximum acetylene concentration in the quenched gas was found to depend on the temperature and C/H ratio of the system: At 3200 K, with a C/H ratio of 0.25 (as for CH_4) the maximum volume percentage of acetylene in the quenched gas is 12; with a C/H ratio of 0.50, the amount of acetylene can be increased to 19%.

3

At 4300 K and a C/H ratio of 15, the maximum concentration is 50%, a value considerably higher than those found in experiments using high intensity arc reactors.

The kinetic aspect of the reaction between graphite and hydrogen has been studied[6]. At temperatures above 3000 K, the sublimation of graphite is the controlling factor, and the reaction rate is independent of hydrogen pressure provided that there are sufficient hydrogen molecules to react with all the C_n species that evaporate. Below 3000 K, the reaction of graphite and hydrogen between 0.01 and 1 atm is a surface reaction whose rate is proportional to the hydrogen pressure and the square root of the dissociation constant of hydrogen. Several authors have investigated the reaction of graphite in low pressure discharges at relatively low temperature of hydrogen. Veprek and coworkers[7] have shown that both the diffusion of H atoms towards the carbon surface and the diffusion of reaction products from the surface are much faster than the rate of surface recombination and the surface reaction. For pyrolytical graphite the probability of the reaction defined as the ratio of the number of C atoms leaving the surface to the number of H atoms impinging on the surface is about 10^{-4} or less. Since the reaction probability would depend strongly on the quality of the carbon used, it would be much higher for carbon of a poor quality, such as that found in coal.

Under plasma conditions any oxygen present in the coal will be evolved as carbon monoxide. If the carbonization is carried out in nitrogen atmosphere, acetylene and hydrogen cyanide will be the main products. Very little cyanogen would be formed unless the nitrogen is greatly in excess of the hydrogen in coal[8]. For a discussion of the thermodynamics of the C–H–N system the reader is referred to an article by Timmins and Ammann[9]. The chemical evaporation, transportation and deposition of carbon in low pressure discharges of oxygen, nitrogen and hydrogen have been described recently[10].

3 Natural Gas and Methane Plasmas

Natural gas as obtained from underground deposits generally has a composition that is significantly different from that of the familiar commercial fuel. The crude gas usually contains some undesirable impurities such as water vapor, hydrogen sulfides, and thiols or other organic sulfur compounds in addition to some heavy, condensable hydrocarbons[11]. Appropriate processing eliminates or reduces the amount of the undesirable impurities and allows the condensable hydrocarbons to be collected as a separate fraction of industrial value. The following volume composition for the commercial fuel is thus arrived[12]: 80–95% CH_4, 8–2% C_2H_6, 3–1% C_3H_8, < 1% C_4H_{10}, < 1% C_5H_{12}, 10–0% N_2, <2% CO_2. The concentrations of the minor components vary slightly with the source of the gas.

3.1 Low Frequency Discharges

The decomposition of methane in the glow discharge has been investigated for many years. At low pressures ethane was the major product[13]. As the methane pressure was increased, ethylene and acetylene were formed; their concentrations in the product became significant if the reactor was cooled in liquid air[14−16]. In a flow system at atmospheric pressure under conditions of high conversion and relatively high temperatures Wiener and Burton[17] found that the yield of acetylene was quite high.

In the negative glow[18] of a dc discharge in methane at low pressures (0.05−0.3 torr) and low currents (0.1−5 mA) ethane, ethylene and acetylene were found in addition to hydrogen and the nonvolatile cuprene, $(CH)_n$, which appeared mainly on the cathode as a solid. Lowering the temperature of the reactor from 77 K to 63 K greatly increased the amount of ethylene. Smaller amounts of propane, propene, propyne, butane, butene, butadiene and pentene were also found. Their rates of formation increased with increasing discharge current at the expense of the C_2 hydrocarbon products and cuprene. The addition of hydrogen to the methane had little effect on the products. Variation of the inter-electrode separation indicated that the products were not formed uniformly throughout the negative glow.

Recently a movable glow discharge of methane has been investigated mass spectrometrically over the whole column length[19]. The mass spectra showed primary fragment ion of methane and ions from condensation reactions up to m/e = 113. The current of different ions reached a maximum very close to the cathode and varied regularly along the axis in maximum and minimum which were related directly with the striations of the column. By simulating the conditions of the glow discharge in a mass spectrometer with high pressure ion source the same ion-molecule reactions were identified with which it was possible to explain the formation of condensation-type ions in the discharge.

Methane has been decomposed in ac discharges operated using 1−6 kV and 30−70 mA in flow systems at pressures of 1−10 torr[20]. With contact times of 0.05−1.5 s the principal products were acetylene, ethane, ethylene and hydrogen together with some higher molecular weight compounds. The conversions which varied from 28 to 91% increased on increasing the contact time and/or discharge current. However, the yields of C_2 and C_3 hydrocarbons reached a maximum at 40−50% total conversion. Almost complete conversions of methane to acetylene have been reported in later works[21, 22]. Vishnevetskii et al.[23−25] have given a set of equations for calculating the rates of acetylene and ethylene formation and the rates of decomposition of several hydrocarbons.

Several attempts[26−31] have been made to analyze the numerous higher molecular weight products from line frequency spark and pulsed discharges in methane at pressures of 250−500 torr. The work is of great interest in connection with the chemistry of primitive earth atmosphere and the origin of life.

The kinetics of conversion of methane to acetylene in glow discharges has been studied in detail[32−35]. A great amount of this work has centered around such parameters as power yields, pressure, cell design, electrode material and presence of hydrogen or argon. Methane conversion was hindered by H_2 or Ar to the same extent and was greater with greater partial pressure of these added gases in the pressure range 40−150 torr. However, the cracking of methane was accelerated by H_2 and Ar at a total pressure of 10 torr. Compared with the higher paraffins, methane yielded the least amount of acetylene. The cracking rate constant increased with pressure. With increasing specific power (power per unit volume of input methane) consumption the degree of conversion increased gradually on inactive and little carbonized, but rapidly on active and carbonized electrodes.

Glow discharges at low pressures in mixtures of methane and nitrogen produced acetylene simultaneously with hydrogen cyanide[36]. C_2H_2/HCN ratios varied according to the current densities and mixture compositions used. When mixtures of methane and carbon dioxide or methane and water were subjected to discharge, carbon monoxide and hydrogen were produced. Acetylene was found only at low current densities[37].

The production of acetylene from natural gas has been studied in a 60 Hz electrical discharge at atmospheric pressure[38]. The apparatus consisted of a cyclonetype reactor in which the products could be removed through a hollow electrode. The electrode separation was 2 cm; the potential difference necessary for spark over gas was 13—16 kV. By varying the residence time of the gas in the reactor from 30 to 600 s a product containing 15—17% C_2H_2 by volume was obtained at relatively low temperatures. Increasing the input rate and decreasing the specific power consumption increased the amount of C_2H_2 formed, but decreased the C_2H_2 concentration in the product. Copper electrodes were reported to produce the least deposition of carbon black and cuprene.

3.2 Triboelectric Discharges

Methane has been converted into ethane, ethylene, acetylene and hydrogen in a "triboelectric" discharge arising from the intermittent contacting of mercury with a glass surface[39, 40]. The discharge is a result of the accumulation of high densities of static charge at the interface by the transfer of electrons from the mercury to the glass. Spectroscopic studies of the discharge have indicated that excited species with energies up to 20 eV above their ground states are present. Further, the triboluminescence spectrum differed from the spark discharge spectrum at atmospheric pressure in that C_2 emission was absent. Both area and nature of the solid surface influenced the extent of breakdown and discharge. The rates of methane conversion were virtually invariant with pressure from 760 to 200 torr, but at 200 torr they increased sharply before gradually falling off again as the pressure was further reduced. The addition of 10% noble gases did not result in any pronounced change in the C_2 hydrocarbon yields or affect the product distribution which was
$C_2H_6:C_2H_4:C_2H_2 = 1:0.34:0.32$.

3.3 High Frequency Discharges

Eremin[41] reported that the amount of methane cracked by a high frequency discharge is proportional to the current consumed and to the amount of excess hydrocarbons. The reaction was found to be of the first order and the rate was directly proportional to the discharge energy and inversely proportional to the original amount of methane. The amount of methane reacting per unit energy consumed was practically independent of the shape of the discharge, the current density and the methane dilution with hydrogen and was lower at higher pressure.

The decomposition of methane has been studied in a 2.45 GHz glow discharge sustained by microwave fields[42]. With 150–310 V/cm, methane pressures in the range 16–220 torr and residence times in the range 0.01–10 s acetylene was the principal product, the maximum yield attained being 31 g kWhr^{-1}. Ethane and ethylene were produced in significant amounts from methane only when the effluent discharge gas was allowed to impinge directly on a liquid-nitrogen-cooled wall. If only stable molecules had been formed in the plasma, the cold wall would not have altered the products. This was taken as evidence that free radicals do not recombine instantly but persist as such for some time after leaving the plasma zone.

Vastola and Wightman[43, 44] found an increase in H_2^+ and a decrease in CH_3^+ when methane was passed through a 2.45 GHz microwave discharge at 0.15 torr and 40 W. With a ten-fold increase in power Baddour and co-workers[45] could efficiently produce carbon black and hydrogen from methane.

Radiofrequency (1–200 MHz) discharges in methane have been studied. Miquel and Chirol[46] observed that the decomposition followed a first-order rate law. Le Goff et al.[47] found that ethane and acetylene constituted 80% of the products. In a recent study Simionescu et al.[48] identified H_2, C_2H_6, C_2H_4 and C_2H_2 among the products of methane decomposition by *in-situ* gas chromatographic analysis. Their ratio was found to depend on the time and power of the rf discharge (Fig. 1). In one investigation[49] products such as ethylacetylene, cyclopentadiene, indene and formaldehyde have also been reported. The formation of formaldehyde appeared to indicate the presence of oxygen or water in the methane sample[50].

Studniarz and Franklin[51] have studied the relative ionic composition of a 50 MHz plasma in methane. Using a quadrupole mass spectrometer and a fast-flow low power discharge they detected CH_3^+, CH_4^+, CH_5^+ and $C_2H_5^+$, in varying proportions depending on pressure, in the range 0.02–0.4 torr. In this work the discharge conditions were chosen so as to minimize polymer formation.

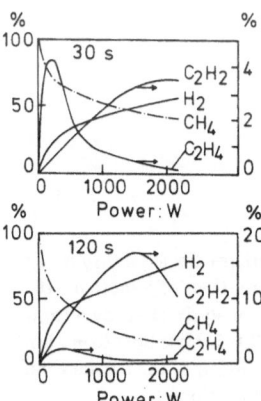

Fig. 1. Power dependence of the decomposition of methane (*broken curve*) and the product yields at 30 s and 120 s discharge periods. (Redrawn from Simionescu, Cr.I., Dumitriu, S., Bulacovschi, V., Onac, D.: Z. Naturforsch. *30b*, 516 (1975), by permission of the publishers, Verlag der Zeitschrift für Naturforschung)

Table 1. Ionic, radical and neutral species observed in an rf plasma in methane[52]

Ions				Radicals	Neutrals	
Major (I > 5%)		Minor (I < 5%)			Major	Minor
High pressure	Low pressure	High pressure	Low pressure			
$C_2H_3^+$	CH_3^+	CH_3^+	H_2^+	$CH_2^.$	H_2	C_3H_4
$C_2H_5^+$	CH_4^+	CH_5^+	H_3^+	$CH_3^.$	C_2H_2	C_3H_8
$C_3H_3^+$	CH_5^+	$C_2H_2^+$	$C_2H_2^+$	$C_2H_3^.$	C_2H_4	
$C_3H_5^+$	$C_2H_3^+$	$C_2H_4^+$	$C_2H_6^+$	$C_2H_5^.$	C_2H_6	
$C_3H_7^+$	$C_2H_4^+$	$C_2H_6^+$	$C_3H_3^+$	$C_3H_7^.$		
	$C_2H_6^+$		$C_3H_7^+$			

With a relatively slow-flow and a 13.56 MHz discharge in methane Smolinsky and Vasile[52] monitored the ionic and neutral products at pressures from 0.1 to 1 torr. Their data are summarized in Table 1. The neutrals were measured by adjusting the potentials on the focusing lenses so as to prevent all ions formed in the discharge from reaching the mass filter. Molecular hydrogen was the dominant product, its mole fraction as well as that of acetylene decreased steadily as the pressure was increased, while those of ethane and ethylene exhibited maxima in the pressure range 0.2–0.6 torr. C_3 hydrocarbons were an order of magnitude smaller than the C_2 hydrocarbons and the higher homologues were proportionately less. C_1 ions (CH_3^+, CH_4^+) and $C_2H_2^+$ decreased monotonically with increasing pressure, perhaps due to reduced electron energy. The change in intensity with increasing pressure of ions such as $C_2H_3^+$, $C_2H_5^+$, $C_3H_3^+$, $C_3H_5^+$ and $C_3H_7^+$ suggested that ion-molecule reactions were occurring in the system. Major differences were found in the observed products when these were sampled axially rather than radially.

Figure 2 shows the different sampling configurations used by Vasile and Smolinsky[53]. Table 2 summarizes the mole fractions of the principal neutral products from

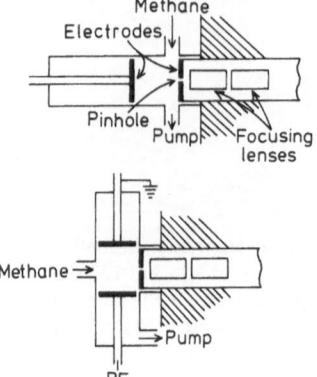

Fig. 2. Schematic representation of the axial (*upper*) and radial (*lower*) sampling configurations of a methane rf plasma for mass spectrometric analysis. (Redrawn from Vasile, M. J., Smolinsky, G.: Int. J. Mass Spectrom. Ion Phys. *18*, 179 (1975), by permission of the authors and the publishers, Elsevier Publishing Company)

Table 2. Mole fractions of neutral products from a 0.45 torr 150 V methane rf plasma[53]

Product	Sampling mode		Radial sampling through an electrically floating orifice
	Axial sampling through the		
	rf electrode	ground electrode	
H_2	0.31	0.23	0.15
CH_4	0.60	0.70	0.78
C_2H_2	0.019	0.013	0.012
C_2H_4	0.013	0.011	0.011
C_2H_6	0.055	0.041	0.045

a 0.45 torr 150 V discharge. Axial sampling through an internal rf capacitor electrode yielded C^+, CH^+, CH_2^+, CH_3^+, $C_2H_2^+$ and $C_2H_3^+$ as the dominant ions, with very little abundance of ions containing greater than three carbon atoms. Radial sampling of a capacitively coupled discharge showed CH_3^+, CH_5^+, $C_2H_3^+$ and $C_2H_5^+$ as dominant with a significant fraction of the total comprising C_4 to C_6 ions. The relative abundances of the C_2-ions are compared in Table 3. The dissimilarity is attributed to the difference in energy of the electrons which cause the ionization and the difference in kinetic energy of the ions that undergo ion-molecular reactions to yield secondary ions.

Investigations[54, 55] of the cracking of natural gas (96% CH_4) by an electrodeless high frequency discharge have been reported. The products were H_2, C_2H_2, C_2H_4, C_2H_6, C_3H_6 and C_3H_8. An increase in the total gas conversion was observed with increasing temperature in the range 220–550 K and contact time in the range 0.04–2.4 s. Addition of H_2 (up to 50% of the total CH_4) caused a decrease, while addition of Ar (up to 75% of the total CH_4) resulted in an increase in the cracking of methane. The effects of the introduction of a number of solid catalysts into the plasma zone have also been reported[55].

Table 3. Relative abundances of C_2-ions produced in a 0.45 torr, 150 V methane rf plasma[53]

Ion	Sampling mode	
	Axial sampling through the rf electrode	Radial sampling through an electrically floating orifice
$C_2H_2^+$	0.092	0.014
$C_2H_3^+$	1.0	0.46
$C_2H_4^+$	0.066	0.137
$C_2H_5^+$	0.36	1.0
$C_2H_6^+$	0.013	0.029

3.4 Electrical Arcs and Plasma Jets

When the current density of a luminous discharge exceeds a certain limit, the potential difference between the electrodes diminishes and the discharge becomes an arc. The arc is characterized by the low potential difference of tens of volts between the electrodes and the high current density which may reach several thousand amperes per square centimeter on the electrodes. When an arc is established between two electrodes in a stream of rapidly flowing gas, the plasma is pushed in the direction of flow. If the arc is formed in a chamber with a proper exit nozzle, a plasma jet is produced outside the chamber with electrons and ions present, in the complete absence of an external electric field.

Following the invention of the high intensity electric arc several attempts have been made to convert methane and natural gas to acetylene. Unlike solids, gaseous reactants induce instability in the arc column by appreciable forced convection[56]. For this reason several arc stabilization techniques were developed. These techniques[57] include confinement of the arc column in a watercooled channel, vortex stabilization, magnetic stabilization, mixing the gaseous reactants with the arc effluent and/or injecting the gases through a specially designed annular nozzle surrounding an electrode.

In their early work Leutner and Stokes[58] used methane-argon mixtures (1:4) in a plasma jet unsuccessfully. These mixtures melted the tungsten cathode instantaneously. In later experiments methane was introduced into the flame of an argon plasma jet. This was accomplished with the apparatus shown in Fig. 3 in which methane was introduced through a water-cooled annulus in the anode normal to the argon jet. About 80% by wt of the methane was converted to acetylene at temperatures estimated to be 1.2×10^4 K. Subsequently Anderson and Case[59] applied thermodynamic and kinetic data to predict the results which could be obtained when methane was introduced into a hydrogen plasma jet and the mixture quenched. The agreement between their analytical treatment and experimental data is very good.

Because of the very high conversion of methane to acetylene the plasma jet process has received more attention than the pyrolytic process. A number of patents describing plasma jet processes of preparing acetylene from methane and higher paraffins have been registered[60-66]. The reported energy consumptions vary from 7 to 11 kWhr kg^{-1} C$_2$H$_2$. The process is considered to be more efficient than the carbide process for manufacturing acetylene[67].

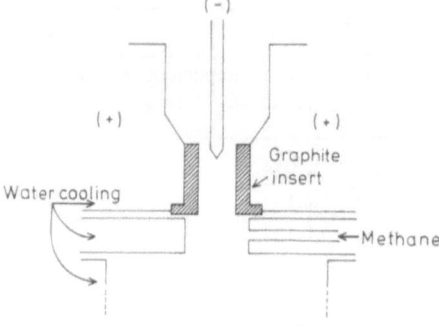

Fig. 3. Argon plasma jet with water-cooled annulus for methane pyrolysis. (Redrawn from Leutner, H. W., Stokes, C. S.: Ind. Eng. Chem. *53*, 341 (1961), by permission of the authors and the publishers, the American Chemical Society)

Eremin and coworkers[68-70] using a high-tension ac arc found that a lower pressure in the 70–800 torr range greatly increased the methane conversion and acetylene production. The power efficiency could be doubled when the pressure was reduced to 70–10 torr. The addition of up to 37% by vol of hydrogen did not affect the acetylene concentration in the reaction products nor the power consumption which was $0.184 \text{ m}^3 \text{ kWhr}^{-1}$. A somewhat lower C_2H_2 yield with a correspondingly higher power consumption was found only with mixtures which contained 48.4% H_2. In pilot-plant experiments the best results were obtained with an initial gas pressure of 42–46 torr natural gas when the power consumption was 3 kWhr m^{-3}. Sixty percent of the methane was cracked and 52% methane cracked with C_2H_2 production. The concentration of C_2H_2 in the product gas was 15%. Current strength and the electrode separation did not affect the proportion of CH_4 cracked or the total power consumption. These results confirmed calculations which were made on the asssumption of a first order reaction in the plasma and which took into account the concurrent reactions $2CH_4 = C_2H_2 + 3H_2$, $2CH_4 = C_2H_4 + 2H_2$ and $CH_4 = C + 2H_2$.

Il'in and Eremin[71-73] reported that the cracking rate in an electrical arc is the same for several paraffins at 1.5 atm pressure and is a function of the energy supplied to the arc. While the maximum concentration of C_2H_2 in the products increased as the molecular weight of the cracking hydrocarbon increased, the energy consumption and heat effects decreased. In each case the degree of overall conversion followed a first order rate law. Preheating the methane to an adequate temperature, for example, 1000 K, increased the yield of C_2H_2 per unit energy input. Decreasing the length of the anode channel did not affect the process significantly but increasing its cross section increased the energy consumption and lowered the volume concentration of acetylene. The addition of $C_3H_8 - C_4H_{10}$ to CH_4 increased the conversion of CH_4 to C_2H_2, the increase being a maximum for 2.5–4 vol % of the additives. It is known that higher hydrocarbons such as propane and butane give good yields of acetylene with less expenditure of energy[74-77]. In a coaxial arc reactor at a specific energy consumption of 3.3 kWhr m^{-3} natural gas conversion into acetylene reached a maximum of 70%.

The effects of preheating natural gas before injection into the plasma jet have been studied by many workers. Whereas Il'in and Eremin[71] observed that preheating at 600 K did not affect the results, Okabayasi et al.[78] found an increase in acetylene yield per unit energy input. However, in agreement with Il'in and Eremin's observation at higher preheating temperatures, Kobozev and Khudyakov[79] reported an increase in C_2H_2 concentration in the products from natural gas preheated to 800 K before injecting it into a 2800 K plasma jet.

In a series of papers and books published over a period of nearly two decades Polak has reported the work of his group at the Petrochemical Research Institute of the U.S.S.R. Academy of Sciences, Moscow[80-88]. An electrode plasmatron was used which operated on a direct current with a maximum voltage output of 15 kV (Fig. 4). Argon containing 10–30 vol % of methane was decomposed at temperatures of 10,000 K and contact times of $10^{-4} - 10^{-5}$ s. Most of the methane was decomposed and 67–87% conversions to acetylene were obtained. The acetylene yield could be increased by using gas mixtures richer in methane[89-91].

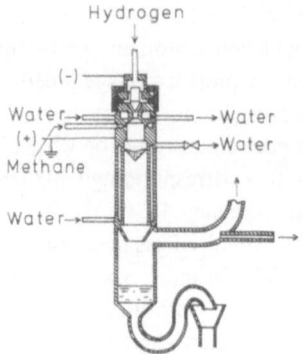

Fig. 4. Schematic of plasmatron system used for methane pyrolysis. (Redrawn from Kinetika i termodinam. khim. reaktsii v nizkotemperaturnoi plazme, Polak, L. (ed.). Moscow: Nauka 1965, by permission of Professor L. Polak, Institute of Petrochemical Synthesis of the Academy of Sciences of the U.S.S.R., Moscow)

Polak's calculations[80] of the kinetics of conversion of methane to acetylene showed that acetylene concentration reached a maximum at a certain distance along the plasma jet (Fig. 5). Chilling $(20-50 \,^{\circ}\mathrm{C} \, \mathrm{cm}^{-1})$ which was begun at a distance slightly less than the distance that corresponded to maximum concentration of C_2H_2, prevented a significant decrease of C_2H_2 in the part of the jet that followed the maximum. Heating of the jet over a short distance of its initial part $(500 \,^{\circ}\mathrm{C} \, \mathrm{cm}^{-1})$ increased the extent of decomposition of methane, raised the maximum concentration of C_2H_2 and displaced the point at which the maximum concentration of C_2H_2 was reached toward the beginning of the jet. The optimum ratio for length to diameter of the reactor was $6.5-7.5$; the number and diameter of the gas (methane) inlets were also significant[87]. In a later work Andreev et al.[92] reported similar variations in product concentrations along their reactor length. Methane pyrolysis in a hydrogen plasma jet with two stage methane addition showed that with proportions and quenching rates suitably adjusted for the second addition, ethylene and hydrogen, and not acetylene, were the major products[88].

With natural gas[83-86] containing 91% CH_4 by vol introduced into a hydrogen plasmatron at 4000–4500 K and with water chilling injected perpendicular to the gas flow the total decomposition was 94% and the conversion to C_2H_2 was as high as 76%. The composition by vol % of a typical gas mixture after the conversion was

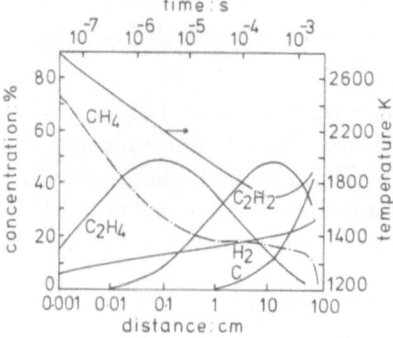

Fig. 5. Relations between the concentrations of methane pyrolysis products and the distance or time over which the reaction occurred. A temperature-time (distance) profile of the plasma jet is also shown. (Redrawn from Kinetika i termodinam. khim. reaktsii v nizkotemperaturnoi plazme, Polak, L. (ed.). Moscow: Nauka 1965, by permission of Professor L. Polak, Institute of Petrochemical Synthesis of the Academy of Sciences of the U.S.S.R., Moscow)

as follows: H_2, 76.5; C_2H_2, 16.5; CH_4, 3.8; N_2, 2.0; C_2H_4, 0.5; C_4H_2, 0.35; small amounts of ethane, propadiene, methylacetylene, vinylacetylene, benzene and naphthalene. Similar results have been reported by Ganz et al.[93]. The energy used for the conversion to C_2H_2 was 45% of the total energy input which is to be compared with the 30–35% used in the production of C_2H_2 by oxidative pyrolysis[81, 83]. The methane to acetylene conversion is similar for the plasma jet pyrolysis and the Du Pont electric arc cracking of methane (80%), but lower for the Huels arc process (50%)[81, 83]. When a combination method of heating natural gas in a plasmatron prior to the hydrogen plasma jet pyrolysis was used the product gases contained less acetylene and more methane indicating a lower conversion rate[94].

The plasma pyrolysis of natural gas with two-stage quenching has also been studied[95]. Methane decompositions at first stage end temperatures of 1700–2000 K and at second stage end temperatures of 110–600 K were considered for various amounts of the plasma-forming hydrogen. Both thermodynamic analysis and experimental results showed that for a given energy input, the two-stage process gave up to 30–40% more unsaturated hydrocarbons than the one-stage process.

Szymanski[96] has investigated the effect of plasma formed in the presence of a ferroelectric sample ($BaTiO_3$ ceramic) on the decomposition of methane. Gaseous end products were similar to those obtained in other types of electrical discharges suggesting similar mechanisms. However, atomic spectra of Ba and Ti were observed indicating that the ferroelectric sample had been activated by the plasma.

Yamamoto[97] has studied the effects of various components in natural gas on acetylene yield. Carbon dioxide decreased the acetylene yield while hydrogen increased the electricity consumption. Nitrogen produced hydrogen cyanide. Dolal[98] generated a mathematical model of the plasma pyrolysis of methane in the presence of steam. Equations were given for calculating the yields and compositions of the pyrolysis products. The plasma-treated natural gas is more efficient than untreated gases in the reduction of iron oxides, especially with increasing temperatures[99].

The plasma pyrolysis of methane in presence of nitrogen has been investigated in detail. Leutner[100] and Stokes et al.[101] used dc arcs and rf devices (24–28 MHz). The dc arc-heated nitrogen mixed with methane gave the best result, both from the standpoint of yield and energy utilization. Increasing the CH_4/N_2 ratio increased the conversion of N_2 to HCN. The rate of quenching and the efficiency of mixing the cold methane with the nitrogen plasma were not determined, but certainly played a role in determining yield and energy efficiency.

Freeman[102–104] used a dc plasma jet to heat a nitrogen gas stream mixed with methane downstream of the arc head. The production of hydrogen cyanide was studied as a function of nitrogen enthalpy, CH_4/N_2 ratio and gas flow rate. At a given enthalpy increasing the CH_4/N_2 ratio increased the HCN yield until a plateau value of CH_4/N_2 was reached beyond which further increases in the CH_4/N_2 ratio did not change the HCN yield. The plateau value of CH_4/N_2 ratio was a function of enthalpy, increasing with increasing enthalpy and leading to increasing yields of HCN. Below the plateau value of the CH_4/N_2 ratio the HCN yield did not appear to be a strong function of enthalpy. The ratio of HCN produced to CH_4 injected is approximately 1/3 at all values below the plateau value. By systematic variation of reactor geometry it was shown[104] that "titration" of either the heat flow or alternatively of nitrogen atomic ions quantitatively accounted for the observed rate of HCN production. Equilibrium calculations[8−10] support these results.

Experiments[105] have been performed with an uncooled rapid transit probe in a nitrogen arc jet in which methane was transformed into HCN and C_2H_2. Locations of chemical reactions were determined and compared with mass spectra of the quenched plasma gas recorded at the same place. The maximum yields of HCN (1.1 mole%) and C_2H_2 (0.5 mole%) were found at a distance of 25 mm along the axis of the jet, but at temperatures of 5500 and 2500 K, respectively.

Bronfin[106] used an induction coupled argon plasma jet to heat methane + nitrogen mixtures. With rapid quenching, the HCN yield expressed as a fraction of the N_2 input converted was found to be a function of input composition and to range up to 70%. In a similar device and in the absence of nitrogen Nishimura et al.[107] found acetylene, hydrogen and solid carbon. When quenched by using a water-cooled silica probe 0.5 cm away from the induction coil the conversion was independent of the power input. However, the selectivity for C_2H_2 formation increased while that for carbon formation decreased with decreasing power input.

Valibekov et al.[108−118] have studied the effect of several parameters on natural gas condensates conversion to acetylene and ethylene in argon and hydrogen plasmas obtained in a variety of arcs and jets operated at 15−30 kW and 3000−5000 K. With condensate fractions boiling at T < 100, 100−200 and >200 °C and pyrolysis temperatures of 1600−2100 K for 10^{-3} to 10^{-4} s the fractional and chemical compositions of the feedstocks did not affect the yield and composition of the product which was mainly acetylene. The conversions varied from 58−95%. Increased residence time reduced the C_2H_2 and C_2H_4 concentration of the product gas, shifted the optimum pyrolysis temperature and decreased the product yield and the degree of conversion. Naturally energy requirements increased. Typically 16−19 vol % of the product gas contained acetylene and 10−13% of C_2H_4 + C_3H_6. In a typical experiment[112], the H_2 consumption was 45 ℓ min^{-1} and the gas condensate consumption 24−88 g min^{-1} at a feed rate in the vapor state of < 3 kg cm^{-2}.

3.5 Laser Irradiation

The decomposition of natural gas by IR radiation of a CO_2 laser was studied in both the presence and absence of C_2H_4[119]. It was found that the rate of decomposition of natural gas was increased by the addition of C_2H_4. The decomposed gas could be used as medium for deposition of thin layers of conductors, semi-conductors and dielectrics of organic and organometallic substances.

4 Petroleum and Petroleum By-product Plasmas

Crude petroleum is a complex mixture containing several different hydrocarbons which are structurally similar. In addition to hydrogen and carbon, small amounts of other elements such as nitrogen (0−0.5%), sulfur (0−6%), oxygen (0−3.5%) and

Table 4. Hydrocarbon fractions from petroleum distillation[120]

Fraction	Composition (range of C atoms)	Boiling point range, °C
Gas	C_1 –C_5	<30
Petroleum ether	C_5 –C_7	30–90
Gasoline	C_5 –C_{12}	40–200
Kerosene	C_{12}–C_{16}	175–275
Gas oil, fuel oil, diesel oil	C_{15}–C_{18}	250–400
Lubricating oils, greases, vaseline	>C_{16}	>350
Paraffin (wax), pitch and tar, petroleum coke	>C_{20}	residue (melts)

some metals in trace amounts have been detected by careful analysis of crude petroleum samples. These additional elements are usually found incorporated into hydrocarbon-like molecules rather than in the free state. The elemental composition of petroleum, particularly the amount of sulfur, is important when the environmental impact of petroleum burning is considered.

Usually crude petroleum is separated into various useful fractions as given in Table 4 [120]. Extensive analyses of crude petroleum have shown that there are more than two hundred hydrocarbons, about one-half of which are low-boiling. Of the 175 hydrocarbons isolated from a representative U.S. petroleum sample, 70 can be classified as paraffins, 48 as cycloparaffins and 57 as aromatic. According to Rossini[121] the hydrocarbons which occur in abundance greater than 1% by vol are: n-hexane (1.8%), n-heptane (2.3%), n-octane (1.9%), n-nonane (1.8%), n-decane (1.8%), n-undecane (1.6%), n-dodecane (1.4%), n-tridecane (1.2%), n-tetradecane (1.0%), methylcyclohexane (1.6%) and 1-methyl-3-isopropylbenzene (1.1%) Petroleums from different locations contain essentially the same hydrocarbons, but the proportions in which the different molecules occur vary considerably.

The behavior of several hydrocarbon fractions from petroleum has been investigated under a variety of electrical discharge conditions. An effort has been made to pyrolyze petroleum or the various hydrocarbon fractions from petroleum distillation. A great deal of this work involves the use of electrode discharges in liquid hydrocarbons.

4.1 Low Frequency Discharges

Early work[122, 123] with C_5–C_{14} paraffins used glow discharges with and without electrodes at low gas pressures. No quantitative analyses were made of the gaseous products which consisted mainly of hydrogen and lower paraffins and olefins. Invariably polymeric solids were produced. In one of the early attempts[124] to analyze the products by gas chromatography an electrodeless discharge of the Siemens type in n-decane was investigated. The major (96.4%) product was hydrogen, the remainder consisted of C_1–C_4 paraffins. Brooks and Hesp[125] have studied the decompo-

15

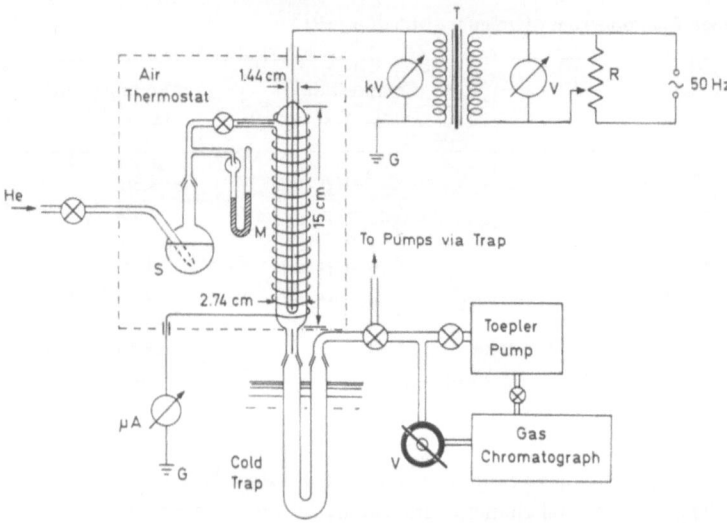

Fig. 6. Diagram of the low-frequency apparatus for decomposition of gasoline and in situ gas chromatographic assembly for product analysis. (Reproduced from Venugopalan, M., Scott, T. W.; Z. physik. Chem. Neue Folge, *108*, 157 (1977), by permission of the authors and the publishers, Akademische Verlagsgesellschaft)

sition of n-heptane vapor in a H_2 or Ar carrier gas using a 50 Hz glow discharge. At 15 torr and 200 W, the production of acetylene was 45%.

In this laboratory the apparatus shown in Fig. 6 was used for the low frequency decomposition of representative samples of U.S. gasoline[126]. Operating at 60 Hz voltages up to 25 kV and discharge currents up to 10 mA significant amounts of methane, ethane, acetylene, ethylene and hydrogen were obtained[127]. With increasing power input polymer deposition on the high tension electrode increased. There was a simultaneous increase in acetylene and methane yields and a decrease in ethylene yields. Fractionation of the feedstock during a run made it difficult to relate the results quantitatively. Attempts to obtain a high degree of conversion to methane by varying the electrical and flow parameters were unsuccessful. The conversion to methane was at best 15% of the input gasoline.

Rowland[128] has claimed that the cracking of low-grade oils by silent corona formed a good-quality gasoline with anti-knock properties.

4.2 High Frequency Discharges

There have been few investigations using high frequency discharges. Coates[129] studied the dissociation of n-hexane by a 2.45 GHz microwave cavity discharge at 1 torr pressure in a flow system. The dissociated products were collected in a cold trap 4 or 77 K) and analyzed by gas chromatography. Of the 25 products detected there was a relatively large number of C_6–C_8 branched chain hydrocarbons. In this work and in a later work[130] it was reported that hydrogen-saturated solid films were

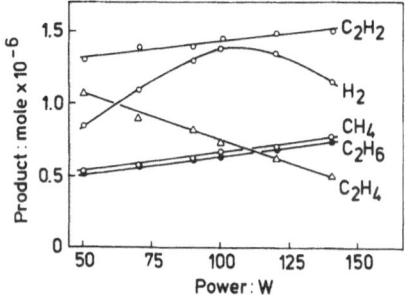

Fig. 7. Distribution of gaseous products as a function of rf power in a n-octane plasma. (Redrawn from Venugopalan, M., Hsu, P. Y.-W.: Z. physik. Chem. Neue Folge, *102*, 127 (1976), by permission of the authors and the publishers, Akademische Verlagsgesellschaft)

produced. However, the decomposition of n-octane vapor in a 13.6 MHz rf discharge-flow system at 0.15 torr produced an unsaturated solid film[131]. Formation of highly cross-linked polymers have previously been reported using electrode discharges in several paraffins[132, 133].

At an n-octane vapor flow rate of 2.14 g/h about 66% decomposition occurred; of the octane decomposed, 15–20 wt% was converted to the solid polymer and the remaining to gaseous products of which H_2 and C_2H_2 were most abundant[131]. Increasing the rf discharge power from 50 to 140 W increased the polymer formation. The product distribution determined by in situ gas chromatographic analysis is shown in Fig. 7. Clearly dehydrogenation – hydrogenation and polymerization reactions occurred in the plasma.

4.3 Electrical Arcs and Plasma Jets

A number of arc reactors and plasma jets have been described for the electrocracking of liquid hydrocarbon fractions from petroleum distillation. Because of higher content of unsaturated hydrocarbons and the absence of O_2, CO and CO_2, plasma pyrolysis is preferred over oxidative pyrolysis.

Il'in and Eremin[134] carried out the pyrolysis of n-hexane, n-heptane, n-octane, iso-octane and low-octane gasoline using a 9–18 kW electric arc at 5000 K in a flowing hydrogen plasma. At gas pressures of 1–2 atm, arc currents of 5–10 A and hydrogen and hydrocarbon consumptions varying in the ranges 9–30 m^3/h and 1–19 ℓ/h, respectively, conversion to gases was nearly complete. The amount of solid products was negligible and 54% of the arc energy was utilized for C_2H_2 and C_2H_4 formation. The data for iso-octane suggested that branching of the molecules did not favor acetylene formation. When a gasoline vapor – superheated (570 K) steam plasma was used for the same gasoline conversion, the yields of C_2H_2 and C_2H_4 were lower than those with the hydrogen plasma.

In the gas phase above petroleum oil fractions at reduced pressures, the products formed were dependent more on gas pressure and type of discharge than on the type of oil[135]. At pressures of 100 torr in an arc, hydrogen and carbon were the main products. As the pressure was reduced, acetylene was formed and at 12 torr the amount of acetylene reached a maximum value which was 22% of the products. In a glow discharge methane, ethylene and other olefins were formed, the maximum value of ethylene being 22% of the products.

Table 5. Plasma jet pyrolysis of petroleum fractions[80]

Material	Pyrolysis temperature	Flow rate			Degree of conversion	Product yield, wt % of reacted material				
		ℓ/min		g/min						
		Ar	H₂	Feed-stock						
	K	Ar	H₂	stock	%	H₂	CH₄	C₂H₂	C₂H₄	C₃H₆
Gasoline	1000	6	–	9.4	73	4.8	7.7	18.0	41.5	–
	1250	–	30	56.7	75	7.0	11.0	28.4	26.0	26.0
	1500	–	28	25.0	87	12.8	9.2	33.0	19.8	21.8
n-Heptane	1500	10	–	8.2	79	7.0	12.5	16.0	29.5	19.0
n-Butane[a]	1900	28	–	1.2[b]	82	2.5	28.8	15.6	49.4	3.7
	2200	26	–	1.1[b]	92	3.0	48.0	18.5	28.5	2.0

[a] 84.5% n-butane + 8.3% n-propane + 6.5% ethane + 0.7% methane.
[b] ℓ/min.

Several papers[80, 81, 83, 136–138] have been published by Polak and coworkers on the plasma jet pyrolysis of butane, n-heptane and low-octane gasoline. Their data are summarized in Table 5. As the number of carbon atoms in the feedstock increased, the yield of methane decreased. With gasoline a hydrogen plasma produced greater yields of acetylene than an argon plasma. Increasing the plasma temperature decreased ethylene and increased acetylene in the product gases. Propylene, which should not have formed at temperatures greater than 1200 K on the basis of thermodynamic considerations, was invariably present in significant amounts. When heptane or gasoline was sprayed into the jet in the form of liquid droplets, the degree of conversion and the yields of individual products were lowered.

Plasma pyrolysis of other oil refinery products as well as pyrolysis of crude oil have been reported. Conversions of 60–70% to C_2H_2 and of 30–50% to C_2H_4 were found for gasoline and diesel fuels[139, 140]. Crude oil yielded $C_2H_2 + C_2H_4$ up to 80% of the feedstock[141]. In the high conversion experiments hydrogen heated to 4000 K in an electric arc was used as the heat carrier and the cracked gases were quenched using cheap, high-boiling oils. By directing an argon plasma jet into the surface of liquid kerosene, 29% C_2H_2 and 10% C_2H_4 were obtained in the product[142]. With a nitrogen plasma jet the yield of acetylene was only 18%[143]. Products of a sample of crude oil in a nitrogen plasma jet[144] yielded ten fractions in the range 376–838 K, boiling range 50° for each cut. Alkanes decreased with temperature from 72 to 0.8% by wt, but the following increased: naphthenes 12–76.7%, aromatics 15.4–32.5%, unsaturateds 0.4–2.8%. The fractions were identified by column chromatography to contain greater than 90 compounds.

In a nitrogen plasma jet concurrent flow of the plasma and petroleum vapor produced 15.4% C_2H_2 and 23.9% unsaturated compounds[145]. With concurrent flow of the nitrogen plasma and liquid petroleum, 7.8% C_2H_2 and 11.5% unsaturated compounds were obtained. Countercurrent flow of the plasma to liquid petroleum produced 12.8% C_2H_2 and 20.5% unsaturated compounds. The addition of liquid petroleum to a plasma jet containing 50–50 mixture of hydrogen and methane produced 16.8% C_2H_2 and 26% unsaturated compounds. For a given energy input two-stage quenching in a hydrogen plasma jet yielded 30–40% more unsaturated hydrocarbons[95].

Many specially designed arc apparatus have been patented for the plasma treatment of petroleum fractions[146–151]. Some of these permit cracking gas-liquid mixtures or gas-solid mixtures in continuous process with recycling of undecomposed petroleum. In one such device[147] crude oil is mixed with an energy transfer medium (rare gas, alkali metal vapor) and the mixture formed into particles which are fed through the hollow cathode of an electrical arc sustained in the transfer medium.

4.4 Submerged Arcs in Liquid Petroleum

Pechuro and coworkers[152–162] have used both non-stationary and stationary electric arcs for the decomposition of liquid petroleum fractions. Besides acetylene and ethylene some methane was produced. For acetylene production power consumption in a stationary arc was lower than in a non-stationary arc. Because of the high dielectric properties of petroleum fractions the preparation of acetylene by the "electrocracking" of liquid petroleum was limited by the small distance between the electrodes. A low voltage ac arc (3 kW) in gasoline, kerosene and diesel fuel yielded 38–46 wt % acetylene.

To produce a low-voltage non-stationary arc[154] transformers were connected so as to provide a voltage of 60–300 V and graphite electrodes and spherical graphite intermediate current-carrying contacts 2 cm in diameter were submerged in gasoline. With increasing voltage the concentration of acetylenic hydrocarbons decreased from 30.9 to 25.6%, the concentration of hydrogen increased from 56.7 to 62.7% and the concentrations of olefinic (8–10%) and paraffinic (3–4%) hydrocarbons remained almost constant, but the total yield of each of these groups of products increased. From 2.2–3.8 kg of petroleum raw material, one cubic meter of acetylene was produced with an electricity consumption of 9.2 kWhr using 16 intermediate contacts and a voltage of 220 V from a 3-phase power system. In an improved apparatus[155, 156] in which the low voltage arc was stabilized with the help of a high-voltage high frequency discharge the product contained more acetylene and fouling of the electrodes with carbon black was prevented.

In the stationary low voltage ac arc (0.4–4 kW) gasoline was passed into the reaction zone through a bore in the lower hollow electrode and circulated[157]. With increasing circulation rate (0–17 ℓ/min) the C_2H_2 yield increased from 22 to 30% while the yields of CH_4 and C_2H_4 decreased and the power consumption decreased from 15 to 7.6 kWhr m^{-3} of C_2H_2. For a given circulation rate an increase in the outer diameter of the hollow electrode and an increase in arc power decreased the hydrogen content of the product gases, increased that of CH_4 and C_2H_4, and did not affect C_2H_2.

In a high voltage ac arc operating at 15 kV and 40 mA the simultaneous decomposition of liquid n-undecane and gaseous natural gas (91% CH_4) hydrocarbons was attempted. The discharge zone increased from 1.25 mm (without added gas) to 5–6.5 mm (with added gas) and consequently the product yields increased[158]. However, the concentration of acetylene decreased with increasing volume of added gas, perhaps due to a dilution effect. When gasoline and oil were cracked in continuous and long runs the products contained 25–27% acetylene[160] and the total conversion of starting material to acetylenic and olefinic hydrocarbons was 52–64%. When one or both of the fixed electrodes were replaced by rotating disk electrodes[159, 161] the arc power, gas yields and rate of carbon black deposition decreased. For acetylene production the power and feedstock consumptions were 8.6 kWhr/m^3 and 2.03 kg/m^3, respectively.

Several patents have been registered for the continuous production of acetylene by intermittent electric arc discharges between a multiplicity of carbon granules immersed between electrodes in petroleum fractions[163–175]. Few articles have reviewed the work from time to time[176, 177]. One design of the apparatus has automatically compensating electrodes for continuous operation[164, 165]. Another design has an electrode arrangement of at least two roller-shaped and rotatable main electrodes and freely moveable auxiliary electrodes supported between the main parallel electrodes[167, 168]. Any change in the form or size of the moveable electrodes changed the power consumption and product output[178]. Furuta et al.[179] have reported that the wear rate of the electrode increased with temperature of heat treatment and that the electrical conductivity of the electrodes was crucial in pulsing the discharge. Addition of a layer of iron sand[166], quartz gravel[170], a water suspension of a hydroxide such as Ca(OH)$_2$, and motion of the conductive coke or graphite granules between electrodes[174] appeared to improve the electrical efficiency. In dc submerged arcs the highest acetylene yields were obtained with Zn, Hg and Pb cathodes, less with Al and least with Fe, Cu and graphite cathodes[180]. Improved acetylene yields were found by using a plurality of arcs and by maintaining the petroleum fraction 15 to 40° below its boiling range at reduced pressure[168, 170, 181].

Selected data on work using submerged arc in liquid petroleum are presented in Table 6. With pulsing the amount of methane in the product gases was increased. Pulsed discharges in liquid hydrocarbons have been studied[171, 172, 182, 183]. With

Table 6. Selected data on work with submerged arcs in liquid petroleum

Petroleum fraction	Arc features	Volume %							Ref.
		H_2	CH_4	C_2H_2	C_2H_4	C_2H_6	C_3H_6	C_3H_8	
n-Propane	60 Hz high voltage	51.2	3.1	30.0	9.5	–	2.5	3.7	148)
n-Butane	60 Hz high voltage	51.2	3.1	30.0	9.5	–	2.5	3.7	148)
Kerosene	60 Hz high voltage	51.2	3.1	30.0	9.5	–	2.5	3.7	148)
	60 Hz low voltage	53.2	6.2	21.5	7.0	–	–	3.0	175)
Fuel oil	60 Hz high voltage	–	–	55	40	–	–	–	166)
Diesel oil	60 Hz low voltage	50	5.0	30	10	–	–	–	170)
Heavy oil	20–100 kHz pulsed high voltage	7.5	15.6	51	7.4	2.4	6.1	0.5	171)
	100 kHz pulsed low voltage	25	11	–	–	45.5	–	–	172)
Mineral oil	60 Hz high voltage	50	–	36	8	4	2	–	182)

heavy oil as the feedstock little ethane was found in one investigation[171] while considerable amount of ethane was reported in another investigation[172]. When iso-octane[182] was used the product yield was independent of liquid flow rate and pulse frequency (200–800 Hz); however, the product yield increased with decreasing pulse time (0.2–1.2 μs) and decreased with increasing electrode separation (0.35–3 mm) and applied voltage (0.6–2.2 kV). For every mole of iso-octane decomposed, 2.19 mole H_2, 1.8 mole CH_4, 1.4 mole C_2H_2, 0.43 mole C_2H_4 and 0.27 mole C_3H_6 were formed. Similar results were obtained for the normal C_5–C_7 paraffins. In another investigation[171] pulsed current increased the acetylene content of the product gases, but considerable amount (20%) of butene was found. Hydrocarbons up to C_5 were found when paraffin oil, transformer oil and so forth were subjected to high voltage discharge[183]. One high voltage pulse discharge technique[184] yielded considerable amount of carbon black (30%) from a number of paraffinic and cyclic hydrocarbons, yet the product gases contained 30% C_2H_2 and 15% C_2H_4[185].

Novikova[186, 187] has studied the micro discharges produced between carbonaceous granules placed in vertical fused silica reactors containing n-heptane or gasoline. Fixed wire electrodes of Mo or Fe were arranged in different planes. Intense arcs were produced when 4–5 mm diamter granules of activated charcoal or bituminous coal coke were used as the moving electrodes. Besides the quality and grain size of the carbonaceous granules the voltage and the number of electrodes influenced the discharge intensity. The following yields of gases were found: 3.5% C_2H_2, 12.4% C_2H_4, 2.2% C_3H_6 and 2.8% C_4H_8 when the electrodes were arranged in a single plane and 8.1% C_2H_2, 19.4% C_2H_4, 3.6% C_3H_6 and 2.7% C_4H_8 when the electrodes were arranged in five planes. Using the latter arrangement for n-heptane and gasoline, the yield of all unsaturated hydrocarbons amounted to 41% and 34.6 wt %, respectively; acetylene amounted to 8.5 and 5.8%, respectively; and ethylene 20.4 and 18.0%, respectively. The concentration of unsaturated hydrocarbons increased with decreasing contact time and increasing voltage, other parameters remaining the same. Increasing the mobility of the particles of carbon packing or the pulse duration in a voltage condenser discharge increased the total yield of unsaturated compounds to 73% and the ethylene yield to 27%[188, 189]. Micro discharges with the presence of significant number of high temperature discharges in the reaction zone, with a relatively low wall temperature and with circulation of kerosene produced 55% C_2H_2, 14% C_2H_4 and 27% H_2[190].

Water emulsions[191–193] of hydrocarbons have been treated in electrical discharges. Water-crude oil emulsions containing 17% water was cracked in a micro discharge yielding more C_2H_2 than an emulsion containing 30% H_2O. The product gases contained 23–28% C_2H_2, 4–6% CH_4 and 5–7% C_2H_4. The power consumption was 9–9.5 kWhr/kg C_2H_2. With olefin foams[194] more polymers than gases were produced.

When the C_2H_2/C_2H_4 ratio in the product gas was high more carbon black separated suggesting that the primary decomposition product was C_2H_2[195]. The deposition of carbon was favored by using an electrode of low heat conductivity such as carbon and by using a feedstock such as decalin, the vapor pressure of which presumably hindered the cooling of acetylene. An electrocarbonizer[196] of liquid hydrocarbons has been described for production of very finely divided carbon. It uses electri-

Table 7. Laser irradiation of crude oil[197]

Product, vol%	Crude oil	
	Vacuum	Helium
H_2	19	49
CH_4	12	11
C_2H_2	10	20
Other hydrocarbons	56	10
$CO + CO_2$	1	9

cal discharges at 40–220 V through moving granules of electrode coke and the liquid which is kept at < 300 K by water cooling.

4.5 Laser Irradiation

Laser pyrolysis with six pulses of 6-J ruby laser of crude oil (39% C, 58.7% H) produced simple gas mixtures of H_2, CH_4, C_2H_2 and higher hydrocarbons[197]. In a helium atmosphere (600 torr) the yield of H_2 and C_2H_2 increased at the expense of higher hydrocarbons while that of CH_4 remained unaffected (Table 7).

4.6 Plasma Desulfurization of Petroleum

The possibility of achieving desulfurization of heavy fuel oils and petroleum products by electrical discharge has been discussed occasionally[197–200]. Among the frequently occurring sulfur compounds associated with petroleum products are the sulfides, disulfides, mercaptans and thiophenes[201, 202].

As early as 1876 Berthelot[203] introduced the vapors of organic sulfur compounds into electric spark and obtained decomposition to C, S, H_2S, C_2H_2 and H_2. Similar treatment of dimethyl sulfide in an ozonizer[204] gave an insoluble liquid containing $C_5H_{12}S_4$, $C_7H_{14}S_6$ and $C_7H_{16}S_5$, in addition to gaseous products which were not analyzed. In recent literature there are few papers which attempt to establish the plasma chemistry of sulfur-containing organic compounds.

Allt and coworkers[199] studied the effects of a low frequency (50 Hz) semi-corona discharge on CS_2, C_2H_5SH, $C_5H_{11}SH$, C_4H_4S and $(C_6H_5CH_2)_2S$ present as 1–2% sulfur-hydrocarbon mixtures. The discharge tube consisted of a quartz air condenser inserted vertically into a 100 cm^3 round bottom flask. The inner electrode consisted of either a stainless-steel or a copper rod passing vertically and centrally through the air condenser; the outer electrode was a thin layer of aluminum foil wrapped around the exterior surface of the condenser. The gap between the inner electrode and the inner walls of the condenser was 3 mm. The discharge with a current of 18 mA was established at atmospheric pressure by applying 15 kV. Sulfur compounds in petroleum spirit were discharged under (1) distillation and (2) reflux

Table 8. Effect of a semi-corona discharge on organic sulfur/hydrocarbon mixtures[199]

Sulfur compound	B. P. range of hydrocarbon in the mixture, $°C$	Effective desulfurization, %	
		Distillation	Reflux[a]
CS_2	40– 60	<0.05	7.3 (17.4)
C_2H_5SH	40– 60	0.3	19.2 (42.8)
$C_5H_{11}SH$	100–120	4.2	3.5 (11.8)
C_4H_4S	80–100	7.0	10.7 (18.6)
$(C_6H_5CH_2)_2S$	156–170	12.9[b]	–
Kerosene	163–259	–	(5–6)

[a] Values given in paranthesis refer to copper electrodes.
[b] Thermal decomposition occurred.

conditions. The extent of desulfurization was followed by direct microchemical analysis and/or gas-liquid chromatography of the various liquid fractions collected at 5 min intervals and of the residual liquors. In almost all instances the reactions were accompanied by the formation of polymeric solid material which varied in color from black for pure CS_2 to yellow for thiophene – hydrocarbon mixtures. Data given in Table 8 show that the discharge experiments performed under reflux conditions yielded better desulfurization results than experiments performed under distillation conditions.

A sample of kerosene containing 0.1 wt% sulfur showed little desulfurization even under reflux conditions[199]. The extent of desulfurization could apparently be extended to as much as 50% by passing the discharged liquid over alumina.

Since hydrogen sulfide was evolved in the electrical discharge, the dissociation of both the hydrocarbon phase and the sulfur compound must have occurred. The absence of H_2S when pure thiophene was used is suggestive of the need for excess H atoms for effective plasma desulfurization.

The plasma chemistry of carbon disulfide is interesting. Invariably a brownish-black solid is formed immediately on initiating the discharge. The solid is believed to be a polymer, probably of composition $(CS)_n$ or $(CS_2)_n$, the former necessarily involving desulfurization. Klabunde and Skell[205] have reported that arc-generated carbon atoms abstract sulfur from carbon disulfide and thioethers yielding $(CS)_n$ polymers. In high-frequency low temperature plasmas CS_2 was readily formed from carbon and sulfur-containing materials[206].

Some work has been done with thiophene and its derivatives using 27 MHz low pressure (1 torr) glow discharges. Fjeldstad and Undheim[207] have found that sulfur is a major product in the plasma reaction of thiophene; phenylacetylene and sulfur were the only products in the case of benzothiophene. According to Suhr[200] thiophene gave a considerable amount of polymers. By addition of hydrogen or water to the thiophene plasma the polymerization could be suppressed and hydrogen sulfide and hydrocarbons produced. Treatment of tetrahydrothiophene in an rf plasma gave hydrogen sulfide and a mixture of unidentified simple hydrocarbons. Dibenzothiophene could be desulfurized quantitatively. Dibenzyl disulfide eliminated sulfur in high yields as S, H_2S and methyl mercaptan and sulfur-free hydrocarbons remained. However, thioanisole gave a number of different products some of which contained sulfur. With dibenzothiophene 5,5-dioxide Suhr and Henne[208] reported 46% conversion to biphenylene and dibenzofuran in the ratio 5 : 2, along with small amounts of dibenzothiophene and naphthalene.

Plasma reactions in which H_2S was eliminated from compounds such as $C_7H_{15}SH$ and $(C_7H_{15})_2S$ have been reported[209]:

$$C_7H_{15}SH = C_7H_{14} + C_7H_{16}$$

$$(C_7H_{15})_2S = C_7H_{14} + C_7H_{16}$$

The selectivity of these reactions is achieved only in low pressure experiments, typically at 1 torr. Using an apparatus similar to that shown in Fig. 6 the elimination of both S and H_2S in the decomposition of dimethyl sulfide has been observed over a wide range of pressures (10–500 torr)[210]. The hydrocarbon products were C_2H_6, C_2H_4 and CH_4, in order of decreasing yield. Both H_2S elimination and methane formation were favored by addition of hydrogen to the discharge. For this reason it may be interesting to study the chemistry of sulfur-containing compounds in hydrogen plasmas. At present sulfur removal from petroleum is usually accomplished by hydrogen treatment at 500–700 K and 3–70 bar in the presence of cobalt and molybdenum catalysts.

When gasoline was cracked in a high-frequency discharge desulfurization was observed with an efficiency of 60%[211]. Very short exposure of heavy oil to a weak discharge resulted in 50% desulfurization[212]. Heavy oils containing dispersed carbon black were desulfurized by exposing to microwave discharges under a hydrogen atmosphere[213]. When a pulsed electric discharge at 100 kHz was used 20% of the original (3%) sulfur in heavy oil was evolved as hydrogen sulfide[214]. Even asphalt has been reported to desulfurize in a hydrogen plasma jet[215].

In view of the recent successful desulfurization of gasoline by metallic sodium[216] at 500–600 K experiments using plasmas in mixtures of sodium vapor and sulfur compounds may demonstrate plasma chemistry that is useful for petroleum desulfurization.

While desulfurization of petroleum fractions is a possibility more quantitative work is necessary before the plasma technique can be applied on industrial scale.

5 Plasmas in Coal

Coal is generally classified according to the degree of increased exposure to the metamorphic conditions of heat and pressure. Four broad sequential stages of coal formation are recognized; each stage represents an increase in the rank of the coal. The approximate compositions observed for the various ranks of coal are given in Table 9 on a dry, mineral-free (dmf) basis. Major changes in composition which occur as the carbonization process goes on are the increase in carbon content and the decrease in

Table 9. The chemical composition of coal[217]

Type of coal	Moisture as found, %	Percent, dry, mineral-free basis (dmf)			
		C	H	O	N
Peat	70–90	45–60	3.5–6.8	20–45	0.75–3.0
Lignite (brown)	30–50	60–75	4.5–5.5	17–35	0.75–2.1
Bituminous	1–20	75–92	4.0–5.6	3–20	0.75–2.0
Anthracite	1.5–3.5	92–95	2.9–4.0	2–3	0.50–2.0

oxygen and moisture contents. In addition to those elements given in Table 9 coal contains many other elements, mostly in trace amounts. Of significance is the amount of sulfur and ash when the environmental impact of coal burning is considered.

The molecular structure of coal is not fully understood[218–220]. A representative structure for coal might consist of small condensed-ring clusters, the outer rings of which are bonded to H or chemical groups that readily replace H. In addition, some atoms of O, N, S and various metals are found in some of the rings or between clusters. Oxygen occurs predominantly as phenolic or etheric groups with less amounts of carbonyls, carboxylic acids or esters. Sulfur is found with bonding similar to that of oxygen. Nitrogen occurs predominantly as pyridine or pyrrolic type rings; the presence of amide groups in coal is undefined. Metals are found as salts or associated with porphyrins.

5.1 Low Frequency Discharges

A few investigations have been reported on the treatment of coal in low frequency discharges. In a high voltage hydrogen discharge at 0.9 torr using internal electrodes the rate of methane formation from sub-bituminous coal (70.5% C, 5.1% H, 23.4% O) increased from 0.1 to 0.4 ml g^{-1} min^{-1} as the voltage was increased from 5 to 6 kV at 30 mA[221]. With increasing hydrogen pressure (0.1–1.0 torr) the rate of methane formation increased at first rapidly and then slowly. When two coals with essentially the same H (5%) and H_2O (7–11%) but different amounts of C (70 and 90%) and O (23 and 4%) and surface areas in the ratio 5:1 were used the rate of methane formation was in the ratio 3:2 suggesting that the reaction occurred essentially at the immediately available surfaces. The formation of methane and smaller amounts of higher paraffins was determined by the reaction locale. Coal inside the luminous discharge zone showed zero order kinetics and farther downstream non-zero order kinetics. In the former case all the hydrocarbons except n-butane were generated at substantially constant rates ($CH_4 : C_2H_2 : C_2H_6 : C_3H_8 = 150:4:1:1$). The rate of n-butane formation fell rapidly to zero with increasing discharge time. For the latter situation all hydrocarbon rates diminished with discharge time. Further, ethylene, instead of ethane, was formed during the initial stage and n-butane throughout the discharge period. From these results Sanada and Berkowitz stated that the higher (C_2–C_4) hydrocarbons were derived from non-aromatic "precursor" configuration in coal.

Based on changes in elemental composition, X-ray crystallographic structural studies, IR and NMR spectral characteristics of coal treated in electrode discharges containing hydrogen Kobayashi and Berkowitz[222] concluded that the overall reaction involved a number of effectively competing processes. The primary reaction was methane formation, presumably by reaction of H with existing CH_3 and CH_2 groups on coal surface and subsequent volatilization to the gas phase. Higher hydrocarbon formation was ascribed to secondary processes, perhaps arising from evaporation of substituent chains or rings in coal and their subsequent reactions with excited hydrogen species in the gas phase.

Electrodeless 60 Hz discharges in bituminous coal have been studied[223]. The apparatus was similar to that shown in Fig. 6 except that the coal was packed into the annular space inside the Siemens tube and degassed. In near room temperature experiments using 26 kV and 2.6 mA, the gaseous products pumping out of the plasma zone in time intervals varying in the range of 10–100 min consisted mainly of H_2 and CO, with small amounts of methane. The addition of He or Ar (25 torr) generally decreased the product yields and the conversion of carbon into methane. More methane was produced in experiments in which hydrogen was added to coal which had not previously been subjected to a plasma. The ESR spectrum of plasma-treated coal[224] showed a 2.5-fold increase (10^{18} spins/g) in the spin concentration over an untreated sample suggesting that a free radical mechanism may be involved in the gasification process. Further work[225] has indicated that low power hydrogen discharges in coal under nearly pyrolytic conditions achieved by heating enhanced the methane production considerably.

Cokes prepared from coal by low-temperature plasma treatment have larger surface area than those prepared by normal or rapid heating[226].

5.2 High Frequency Discharges

In the high-frequency category of discharges, several investigations of coal have been reported. All of these investigations use microwave (2.45 GHz) excitation. Use of rf (1–200 MHz) excitation has not been reported yet, perhaps due to difficulty in sustaining the discharge at gas pressures produced on volatilization of the coal.

The microwave discharges were produced at power levels of 25–50 W in an air-cooled Opthose coaxial cavity in which the coal sample was placed. Both static and flow systems were used and the discharge pyrolysis of coal was performed in vacuo and in the presence of added gases such as Ar, H_2, O_2, N_2, CO_2 and H_2O[227, 228]. The types of coal studied included lignite, low (lvb), medium (mvb) and high (hvab) volatile A grade bituminous coal and anthracite (Table 9). The conversion of carbon to hydrocarbons was found to depend on the amount and type of volatile materials, the carbon content and type of carbonaceous materials, as well as the added gas and the type of reactor used.

5.2.1 Static Systems

In a series of papers Fu and Blaustein[229–231] reported that coals of various ranks readily evolved enough of their volatiles to initiate and sustain the discharge at power levels of 50 W. Tars were deposited on the reactor wall immediately after the discharge appeared. Except for lignite (brown coal) for which the pressure rise was quick there existed usually an induction period before an extensive build-up of gases. In presence of argon the induction period for gas evolution did not exist. For each rank, the pressure reached a plateau after a short period, this period being shorter in the presence of argon; the extent of volatilization increased with volatile matter of the coal, but was unaffected by the added argon.

Mass spectrometric analyses of the products from four different vitrains are given in Table 10. The major products were H_2 and CO and the minor products C_2H_2, CH_4 and CO_2. Significant amounts of CO and CO_2 were obtained from lignite due to the presence of relatively large amounts of oxygen in the sample. When the decomposition[227] of lignite was carried out in a CO_2-Ar (4:1) discharge, the hydrocarbon yields decreased, but CO formation increased with time indicating that the hydrocarbons were oxidized by active species produced from CO_2. However, an Ar–H_2 or Ar–H_2O discharge decreased the CO_2 formation while CO formation was little affected. Except in the case of lignite, the addition of H_2 appeared to favor the hydrocarbon production.

For the investigation of composition of the gases evolved at different stages of the volatilization, pyrolyses have been carried out with interruptions of the discharge[231]. For both lignite (67% C) and hvab coal (82% C) the concentration of H_2 increased but that of CO and CO_2 decreased, in analogy to thermal decomposition

Table 10. Products from discharge pyrolysis of coal in vacuum and in the presence of added gases in static systems[230, 231]

Rank of coal	Reaction time, min	Pressure of added gas, torr			Product, mmol/g solid				
		Ar	H_2	H_2O	H_2	CH_4	C_2H_2	CO	CO_2
Lignite	10	–	–	–	8.65	0.27	0.78	8.35	0.87
	5	5.1	–	–	8.68	0.21	1.04	7.95	0.86
	1	2.2	21.6	–	2.54[a]	0.21	0.93	6.62	0.35
	1	8.8	–	11.0	8.9	0.15	1.00	8.72	0.50
lvb	20	–	–	–	12.0	0.14	0.79	1.13	0.01
	20	5.1	–	–	11.3	0.40	0.88	1.08	0.02
	1	2.2	21.6	–	[b]	0.38	1.75	0.81	0.01
	1	7.7	–	10.0	2.09	0.11	1.36	2.58	0.05
hvab	20	–	–	–	10.3	0.35	1.50	3.57	0.08
	20	5.1	–	–	9.85	0.25	1.58	3.19	0.07
	1	2.3	22.7	–	6.39[a]	0.47	3.11	2.11	0.02
	1	8.2	–	10.5	5.64	0.20	2.44	4.48	0.20
Anthracite	20	–	–	–	6.02	0.50	0.18	0.39	c
	20	5.1	–	–	6.45	0.04	0.02	0.42	c
	1	2.4	22.6	–	[b]	0.46	0.96	0.30	c
	1	5.0	–	7.3	4.54	0.06	1.04	4.24	0.08

[a] Net increase in H_2. [b] Net decrease of H_2. [c] Trace amount.

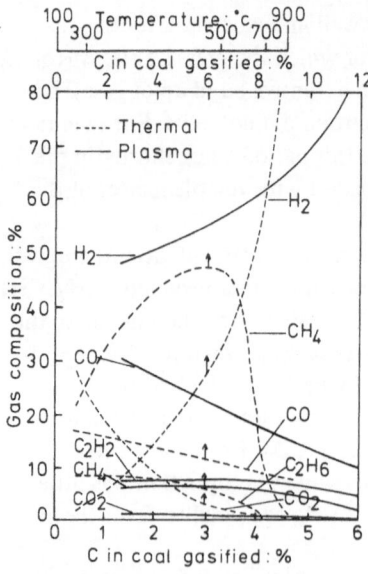

Fig. 8. Gas composition at various stages of microwave pyrolysis (solid curves) and thermal pyrolysis (broken curves) of hvab coal. (Redrawn from Fu, Y. C., Blaustein, B. D.: Ind. Eng. Chem., Process Design Develop. 8, 257 (1969), by permission of the publishers, the American Chemical Society)

of coal where similar trends were observed as the decomposition progressed with temperature (Fig. 8). This indicated that in the later stages the surface oxygen groups were largely removed and the gas evolution was chiefly due to volatilization of H species and perhaps hydrogenated carbon species. The major hydrocarbon products of discharge pyrolysis were C_2H_2 and CH_4 while those of thermal pyrolysis were CH_4 and C_2H_6. In the discharge pyrolysis, the concentrations of the hydrocarbons were nearly constant in the earlier stages; they, however, decreased at the later stages. In the thermal pyrolysis the ethane concentration varied in a similar fashion, but the methane concentration passed through a maximum near 800 K. Certainly, this temperature region of plasmas in coal needs to be further investigated (Sect. 5.1).

The absence of higher hydrocarbons in the products suggested that fragmentation rather than combination of radicals detached from coal was favored under plasma conditions. This aspect was confirmed by cooling one end of a h-shaped reactor with liquid nitrogen (77 K) while the other end containing the hvab coal was discharged. Indeed, substantial amounts of higher hydrocarbons were formed and the yield of gaseous products as well as the conversion to hydrocarbons were increased.

Experiments similar to those described above have been reported for nearly the same conditions but with added N_2[232]. In addition to the usual H_2 and CO, HCN, C_2H_2, small amounts of $(CN)_2$ and CH_4 were found in the products. By trapping at 77 K more than 42% of C in the hvab coal was converted to HCN and C_2H_2. Of course, small amounts of higher hydrocarbons were formed in the cold trap.

5.2.2 Flow Systems

A number of experiments[227, 228] have been performed using reactors (a) and (c) shown in Fig. 9; reactor (b) has the disadvantage of limiting the flow rate because

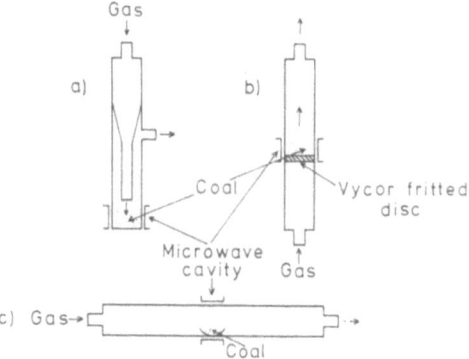

Fig. 9. Reactors used in microwave discharge-flow system for coal gasification. (*a* and *b* redrawn from Fu, Y. C., Blaustein, B. D., Wender, I.: Chem. Eng. Progr. Symp. Ser. *67*, 47 (1971), by permission of the authors and the publishers, American Institute of Chemical Engineers)

of the fritted disk. Most experiments done with reactor (*a*) used a flow rate of 1 cm^3/min at 10 torr pressure. The initial rates of total gas production were estimated to be 4.9×10^{-2} and 2.1×10^{-2} mole min^{-1} g^{-1} for lignite and hvab coal, respectively. Gasification of 10 mg samples was nearly complete in 30 min except for slow degassing of small amounts of H_2. The extent of gasification after 30 min was 42.5% for lignite and 35.5% for hvab coal. Hydrogen and carbon monoxide were the major products; methane, acetylene and ethylene were the major hydrocarbon products.

Variations in rate of production of H_2, CO, CO_2, CH_4 and C_2H_n ($= C_2H_2 + C_2H_4$) from lignite and hvab coal in Ar discharge are shown in Figs. 10 and 11. The initial production rates of methane and the other hydrocarbons were greater with lignite than with hvab coal. Figure 10 also compares variations in the rate of gas production from lignite in Ar and H_2 + Ar (4:1) discharges. The constancy of the rate curve for H_2 in the later stages represented the constant flow of hydrogen as the reactant gas. In the initial stages the rates of formation of CO, CO_2 and C_2H_n were decreased and of CH_4 appeared to be unaffected. The rate curves for methane formation, however, showed that the discharged hydrogen continued to react with the remaining carbon in lignite to form methane at a constant rate in the later stage. When CO_2 was used in place of hydrogen, the rate of hydrocarbon formation was reduced drastically.

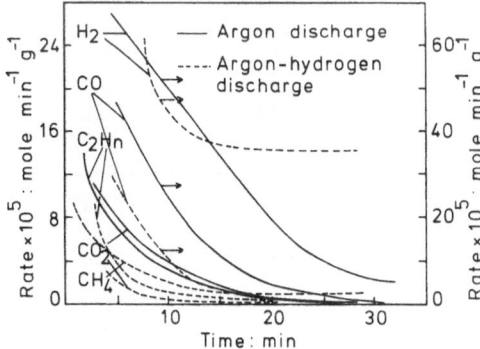

Fig. 10. Variation in rates of H_2, CO, CO_2, CH_4 and C_2H_n ($= C_2H_2 + C_2H_4$) formation from lignite in argon (solid curves) and in argon-hydrogen (broken curves) discharges with time. For both discharges the rates of methane formation are identical. (Redrawn from Fu, Y. C., Blaustein, B. D., Wender, I.: Chem. Eng. Progr. Symp. Ser. *67*, 47 (1971), by permission of the authors and the publishers, American Istitute of Chemical Engineers)

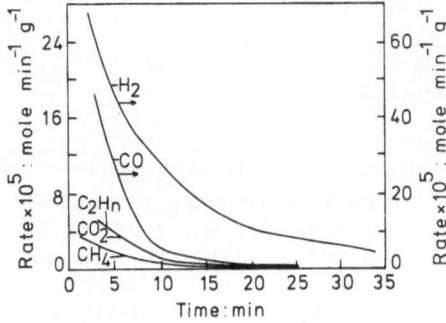

Fig. 11. Variation in rates of H_2, CO, CO_2, CH_4 and C_2H_n (= C_2H_2 + C_2H_4) formation from hvab coal in argon discharge with time. (Redrawn from Fu, Y. C., Blaustein, B. D., Wender, I.: Chem. Eng. Progr. Symp. Ser. *67*, 47 (1971), by permission of the authors and the publishers, American Institute of Chemical Engineers)

When the discharge products were condensed in a liquid nitrogen-cooled trap, acetylene, ethylene, higher molecular weight hydrocarbons (up to C_8, including cyclic compounds), CO_2, HCN and water were detected. Table 11 contains the mass spectrometric analyses of products from discharge flow systems after 30 min of reaction time. The non-condensable gases H_2, CH_4 and CO apparently accounted for the differences between the weight loss of solid and the weight of condensable products collected. A comparison with the data given in Table 10 suggested that ethylene was formed at the expense of acetylene. Losses in weight of the solids in the flow system were higher than the amounts of the total gaseous products obtained in the static system. Perhaps the attainment of a steady state is responsible for the smaller extent of gasification in the static system.

Table 11. Condensable products from discharge-flow systems[227]

| Product | mmol/g solid | | | |
| | Hvab coal | Lignite | | |
	Ar	Ar	Ar + H_2	Ar + CO_2
C_2H_2	0.96	0.75	0.99	0.24
C_2H_4	0.26	0.41	0.21	0.03
C_2H_6	0.09	0.07	0.06	0.02
C_3H_6	0.03	0.04	0.03	0.01
C_3H_8	0.01	0.01	0.01	b
C_4H_2	0.03	0.06	0.05	0.01
C_6H_6	0.02	0.02	0.02	b
CO_2	0.86	1.71	1.46	3.11[a]
HCN	0.37	0.13	0.13	0.06
H_2O	0.49	0.39	0.48	0.37
Total condensables, wt % solid	9.5	12.5	14.8	21.4[a]
Wt. loss of solid, %	35.5	42.5	43.7	51.0

[a] Unreacted CO_2 included.   [b] Trace amount.

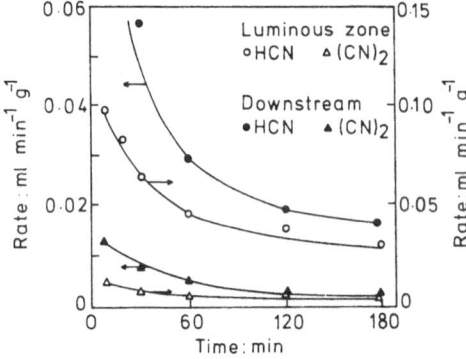

Fig. 12. Variation in rate of HCN and $(CN)_2$ formation from mvb coal in nitrogen discharge with time. Open points for rates in the luminous zone; shaded points for rates downstream from the plasma. (Redrawn from Nishida, S., Berkowitz, N.: Fuel *52*, 262 (1973), by permission of Dr. Berkowitz and the publishers, IPC Business Press Ltd.)

There is only one report[228] on the gasification of coal in a nitrogen discharge-flow system. These experiments were performed with a variety of subbituminous and bituminous Canadian and Japanese coals. Typically a 50 mg sample was evacuated at 400 K and discharged in reactor (*c*) of Fig. 9 with a nitrogen flow of 5 cm³/min at power levels of 25–50 W. Reaction temperature during a run was controlled and measured. Products were analyzed gas chromatographically. The major product was HCN which, depending upon the rank of the coal, was initially produced at rates ranging from 0.5 to 0.03 cm³ g⁻¹ min⁻¹ at STP. As reaction continued, these rates invariably fell quickly, usually dropping to to one-third or less of the initial value after 2–3 h. Generation of cyanogen proceeded much more slowly than hydrogen cyanide in the early stages of reaction. Typical rates were one-fifth to one-tenth of the corresponding rate for HCN.

Neither rates were influenced by the reaction locale, that is, by whether the reaction took place in the luminous discharge zone or some short distance from it (Fig. 12). However, unlike the HCN production rate, the CN rate generally tended to attain constant value after 1 h. In consequence, the $(CN)_2$/HCN yield ratio increased slowly as reaction progressed. For a Canadian coal (88.3% C, 5.1% H) this ratio changed with temperature of the reactor, steadily decreasing up to 460 K and thereafter rising again (Fig. 13). The inset in Fig. 13 shows the manner in which

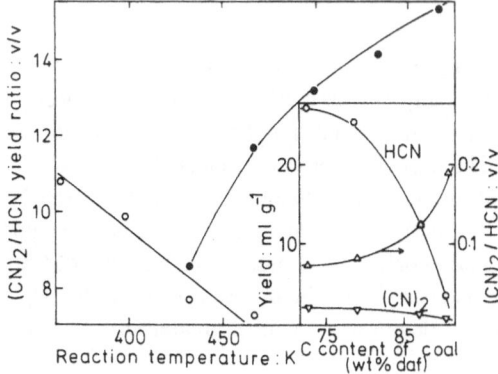

Fig. 13. Variation of $(CN)_2$/HCN yield ratio with temperature for Canadian coal, as determined for 30 min reaction period. Shaded points for reaction in the luminous discharge zone; open points downstream from the luminous zone. *Inset* shows variation of cumulative 2-h yields and $(CN)_2$/HCN yield ratio with rank of Canadian coal. (Redrawn from Nishida, S., Berkowitz, N.: Fuel *52*, 262 (1973), by permission of Dr. Berkowitz and the publishers, IPC Business Press Ltd.)

cumulative 2 h yields of HCN, $(CN)_2$ and the $(CN)_2$/HCN yield ratio varied with rank of coal. Neither rates of production were changed by varying the particle size of coal. Unlike the generation of HCN, $(CN)_2$ and CO_2 the rate of formation of CO was determined by the reaction locale.

Progress of the reactions in the above system was studied by measuring changes in the intensities of several diagnostic IR absorption bands with respect to time. Based on IR spectral changes accompanying the reaction and the rate measurements it was concluded that HCN was mainly formed from non-aromatic CH in coal and that $(CN)_2$ was formed for the most part in aromatic structures, though some could also form in competition with HCN from non-aromatic CH. Rate data obtained in subsequent experiments in which nitrogen-hydrogen mixtures were used supported these conclusions. In these experiments[233] the rates of HCN production were 10–50 times greater than in the nitrogen discharges. The generation of hydrocarbons was minimal and detectable amounts of ammonia were not found. Based on additional experiments[233] in which pure H-free spectroscopic carbon was used in place of coal, it was further concluded that the enhanced formation of HCN was controlled by transient hydrogenation of aromatic carbon and consequent generation of additional nonaromatic reaction centers which could then be abstracted by excited nitrogen atoms.

5.3 Electrical Arcs and Plasma Jets

Both the electric arc and the plasma jet have been used for the pyrolysis of coal. Acetylene is the principal hydrocarbon product, its yield being three times more in a hydrogen atmosphere than in an argon atmosphere. Since the thermodynamic stability of acetylene decreases rapidly below about 1600 K, the product gases must be quenched rapidly in order to prevent the decomposition of acetylene (Sect. 2).

During the last decade several attempts have been made to optimize the pyrolysis process with respect to coal rank, particle size, the discharge atmosphere and quenching.

5.3.1 Electrical Arcs

The early arcs[234–236] used coal as the consumable anode of a direct current arc (Fig. 14). In the process crushed coal, typically 10–20 mesh, was fed into an electric discharge sustained between a graphite cathode and the coal at a feed rate consistent with the surface pyrolysis rate. The rapid heating occurring at the surface pyrolyzed the coal. The hydrocarbon products were quenched downstream of the arc zone by injecting a gas to preserve the acetylene produced in the discharge region; the solid residue of char and undecomposed coal, if any, spilled over the sides of the anode feed tube. Gas chromatographic analyses[236] of the products showed that at power levels of 15–45 kW the yield of acetylene was 2–3 times greater with hydrogen than with argon as the quench medium.

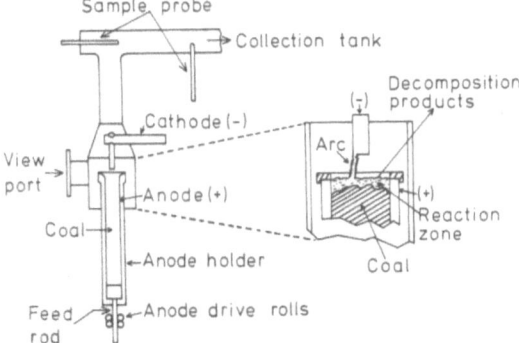

Fig. 14. Coal gasification. Arc coal reactor. (Redrawn from Krukonis, V. J., Gannon, R. E., Modell, M.: in Advances in Chemistry Series No. 131, pages 30–33 (1974), Massey, L. G. (ed.), by permission of the authors and the publishers, the American Chemical Society)

Based on the minimal acetylene yields in experiments[236] using hydrogen-free char in the arc environment it was concluded that additional reactions occurring between carbon in coal and the gaseous hydrogen environment were not responsible for the improved acetylene yields with a hydrogen quench. Further, in the absence of quench the decomposition of acetylene in the product flow occurred readily. One of the most serious causes of acetylene decomposition was the contact between the incandescent surface char layer and the acetylene which was generated below the surface (Fig. 14).

A schematic of a rotating arc[236, 237] is shown in Fig. 15. Powdered coal, typically 100 mesh, was carried downward via hydrogen through a magnetically rotated arc region at a velocity of several m/s. A few cm downstream the hydrocarbon flow was quenched. As in the consumable anode arc, hydrogen produced substantially higher acetylene yields than did an argon quench. When acetylene itself was injected into a C-free plasma stream, it was found that the extent of acetylene decomposed was 60% with an argon quench as compared to 10–12% with a hydrogen quench.

In another experiment[236] it was found that the decomposition of acetylene in both hydrogen and deuterium was identical and low relative to He, Ar or N_2 suggesting that physical properties such as thermal conductivity, diffusivity or heat capacity of the quench were not the reason for the acetylene-preservation in hydrogen. Mass

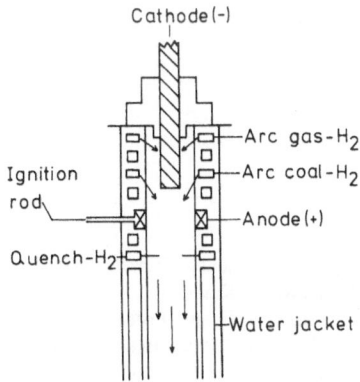

Fig. 15. Coal gasification. Rotating arc coal reactor. (Redrawn from Krukonis, V. J., Gannon, R. E., Modell, M.: in Advances in Chemistry Series No. 131, pages 30–33 (1974), Massey, L. G. (ed.), by permission of the authors and the publishers, the American Chemical Society)

spectrometric analyses of the deuterium samples showed that H-D interaction to form C_2HD (15%) and C_2D_2 (84%) was occurring during the plasma quenching step. This indicated that almost all of the C_2H_2 had exchanged with D_2 to form C_2HD and C_2D_2. Thus, a previously proposed quenching mechanism[5, 7)] which suggested rapid cooling of C_2H_2 and also required recombination of C_2H radical with H appeared unlikely. In yet another experiment[236)] a 50–50 mixture of ^{12}C and ^{13}C acetylenes were injected into a hydrogen plasma stream. Mass spectrometric analysis showed that nearly all the carbons in the acetylene were exchanged.

To account for the essentially complete interchange of both C and H atoms Krukonis et al.[236)] suggested a chain mechanism in which the chain was initiated by the fragmentation of a relatively few C_2H_2 molecules into C_2H, C_2, CH and H species. Collision of the fragments with C_2H_2 molecules then led to exchanging of atoms and splitting to additional fragments. As the reaction mixture cooled downstream, the number of collisions would decrease and the chain would eventually be terminated by the recombination of two CH fragments or a C_2H and H to form acetylene.

Pyrolysis of lignite injected into an argon arc axially through an annulus in the cathode yielded 23 vol % C_2H_2 at 3900 K compared with 10% at 5900 K for coal injected radially into the arc[238)].

Ladner and Wheatley[239)] have described an intermittent submerged arc apparatus with horizontal fixed graphite electrodes operated by 50 Hz current. In a typical experiment the graphite electrodes were immersed in 200–300 cm^3 of liquid coal feedstock prepared by digesting a low-rank coal (82% C, 5.2% H) under pressure in anthracene oil. Hydrogen and acetylene were the major products. The highest concentration of acetylene was produced when the arcs were of short (0.4 ms) duration. The concentration of acetylene decreased with increase in coal content of the feedstock. It was estimated that the average consumption of the system was 7 kWhr/kg C_2H_2.

5.3.2 Plasma Jets

Various plasma jet devices have been used to study the discharge pyrolysis of coal. One such device used by a number of workers is shown in Fig. 16 [240–245)]. A direct

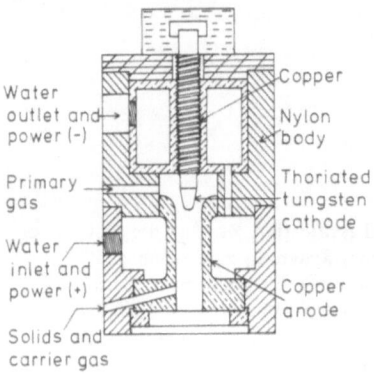

Fig. 16. Plasma jet used in coal gasification. (Redrawn from Graves, R. D., Kawa, W., Hiteshue, R. W.: Ind. Eng. Chem., Process Design Develop. 5, 59 (1966), by permission of the publishers, the American Chemical Society)

Table 12. Product yields and carbon conversion in different atmospheres

Coal injection	Coal feed-rate, g/min	Plasma gas[a] vol%	Total power input, kW	Carbon conversion, wt %			Ref.
				C_2H_2	HCN	CH_4	
Radial	0.08	Ar	4.2	18	–	–	241)
	5.2	Ar	20	52	–	–	244)
	0.08	H_2	4.3	18	–	–	241)
	5.2	H_2	11	65	–	–	245)
	0.05	92% Ar + 8% H_2	4.2	37	–	–	242)
	0.06	90% Ar + 10% N_2	7.0	3.4	12.8	0.8	242)
	0.31	N_2	13	2.7	19.0	0.1	242)
Axial	0.05	Ar	7	37	–	–	253)
	0.05	90% Ar + 10% H_2	13.6	74	–	–	253)

[a] Flow rate = 30 ℓ/min.

current arc was struck between a thoriated tungsten cathode and a tubular copper anode, both of which were water cooled. The primary gas, called arc gas, consisting of Ar, H_2 or N_2 was passed through the electrode space. The coal samples were injected radially into the plasma by suspending these in a carrier gas, usually a stream of Ar using a fluidized bed. Recently a water-stabilized plasma generator combined with a jacketed vertical tube pyrolyzer was described for the high temperature pyrolysis of coal[246].

The plasma jet reaction of coal invariably produced gas and finely divided soot. Tarry material was not found. With argon or hydrogen plasma jets, the main gaseous products were H_2, CO and C_2H_2; with nitrogen plasmas HCN largely replaced C_2H_2. The percent conversion of carbon in coal to the various products for different arc gases is given in Table 12. The data show that the conversion to acetylene is increased by axial injection of coal and by adding hydrogen to the plasma. In obvious disagreement with the data of Bond et al.[241, 242] Kawana and coworkers[244, 245] found that a H_2 plasma gave twice the C_2H_2 yield (15 g/kWhr) obtained in an Ar plasma. However, in agreement with the data of Bond et al. Kulczycka[247] obtained in a H_2 + Ar

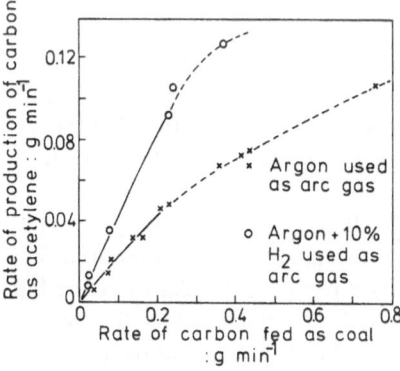

Fig. 17. Effect of coal feed rate on conversion to acetylene. The rectilinear portions of the graph, represented by the solid lines, correspond to a 20% conversion of C in the coal to C in C_2H_2 in Ar plasma and to 40% conversion to C_2H_2 in the 90% Ar + 10% H_2 plasma. (Redrawn from Bond, R. L., Ladner, W. R., McConnell, G. I. T.: in Advances in Chemistry Series No. 55, page 659 (1966), Gould, R. (ed.), by permission of Dr. Ladner and the publishers, the American Chemical Society)

35

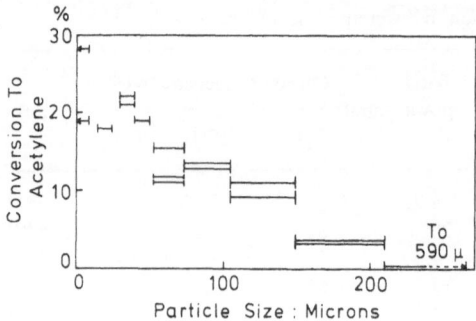

Fig. 18. Effect of coal particle size on the percentage conversion of carbon in coal to acetylene in an argon plasma. (Redrawn from Bond, R. L., Ladner, W. R., McConnell, G. I. T.: in Advances in Chemistry Series No. 55, page 659 (1966), Gould, R. (ed.), by permission of Dr. Ladner and the publishers, the American Chemical Society)

plasma twice the C_2H_2 yield found in an Ar plasma. A recent patent[248] describes a process in which lignite dust injected into H_2 plasma jet for 1 ms produced a gas containing 52% H_2, 25% CO and 19% C_2H_2. Srivastava et al.[249] have given an equation for the molar free energy function for 11 species produced in a coal plasma jet, from which theoretical yields of products may be calculated.

The effects of coal feed rate and particle size on conversion to acetylene in an Ar plasma are shown in Figs. 17 and 18. The rectilinear portions of the graphs in Fig. 17, represented by the solid lines, corresponded to a 20% conversion of the carbon in the coal to carbon in acetylene in an argon plasma and to 40% conversion to acetylene in 90% Ar + 10% H_2 plasma. At coal feed rates exceeding about 0.3 g/min the yield of acetylene diminished. This could be caused by either inefficient heating of the solid feed material or inadequate quenching of the gaseous products, or both. The extent of the reaction strongly depended upon the particle size, particles with diameters greater than about 200 μ seldom reacting. Figure 18 shows that for particles of 50 μ or less in diameter a 20% yield of acetylene was reached, which was the yield obtained for the coal ground to pass a 70 mesh sieve. Similar results have been reported by Kulczycka[247, 250] who worked with Polish coals.

The yields of C_2H_2 in argon plasmas and of C_2H_2 and HCN in nitrogen plasmas increased linearly with increasing amount of volatile matter in the coals[240, 242]. According to one investigation, rankwise, anthracite gave 5% C_2H_2, lignite 11–18% C_2H_2 and bituminous coal 14–24% C_2H_2[250]. Indian coals which contained more ash than British or American coals gave the same yields[251]. The high mineral-matter content of coals increased the energy required per unit of acetylene produced.

Yields of gaseous products increased with increasing plasma temperature (Table 13)[243]. In all experiments, plasmas contained sufficient energy to heat the coal to temperatures considerably higher than needed for complet devolatilization, but the solid residues contained 10% or more volatile matter, indicating that thermal equilibrium between the coal and plasma was not reached. In another investigation similar results have been reported[250].

Littlewood[252, 253] has made several modifications to the Ar plasma jet so that the pyrolysis efficiency with respect to acetylene formation could be improved. To ensure that the coal was introduced into the hottest region of the plasma, it was fed axially into the plasma through holes drilled in a hollow cathode holder. The principal argon stream was introduced tangentially into the plasma jet and the pyrolysis

Table 13. Effect of plasma jet temperature on product yields[243)]

Average plasma temp., K	Average coal feed rate, g/min	Net power input, kW	Solid residue %	Gas product, wt % maf coal			
				H_2	CH_4	C_2H_2	CO
		70 x 100 mesh					
4200	9.1	2.0	87.7	0.4	0.2	2.2	3.8
6900	7.5	3.3	84.0	1.0	0.1	4.3	5.6
8900	8.4	4.7	78.3	1.7	0.2	6.0	11.0
		−325 mesh					
5100	7.8	2.4	73.6	2.4	2.7	9.5	18.1
7600	6.4	3.7	62.9	3.0	0.5	12.3	11.4
9100	5.6	4.9	45.3	3.9	0.6	15.4	24.3

products were then passed through a reaction tube for quenching. Using a bituminous coal sample with 36% volatile matter the maximum yield of C_2H_2 in an argon plasma was equivalent to a conversion of 37% by wt of C in the coal to C_2H_2 (Table 12). When Ar + H_2 mixture was used as the principal arc gas a maximum C_2H_2 yield of 74% by wt on a C basis was found (Table 12). This is the highest yield of acetylene from coal yet reported. It appears that the penetration and hydrogenation of the coal-carrier gas stream, which was injected axially rather than radially into the plasma stream is a critical factor in the acetylene production.

For the pyrolysis mechanism it is considered that acetylene is formed principally from the aliphatic and alicyclic components of coal structure. This in turn is possibly related to the volatile material content of coal. Nicholson and Littlewood[253)] tested this hypothesis using coal tar pitch and methane as feedstocks. The maximum yield of C_2H_2 from pitch was only 35% on a C basis, chiefly because of its highly condensed aromatic structure, whereas the yield from methane was 88%. The cracking of aliphatic and alicyclic fragments approximating to the general formula C_nH_{2n} to form acetylene as the ultimate product involves dehydrogenation steps resulting in the appearance of at least one molecule of hydrogen for each molecule of acetylene formed. This imposes an upper limit on the yield of acetylene from a given coal. In Nicholson and Littlewood's experiments the maximum yields obtained for a number of coals were all very close to their limiting values. Possible means of increasing the limiting values are addition of more hydrogen to the coal structure be hydrogenation before pyrolysis or pyrolysis of coal in the presence of a significant partial pressure of hydrogen. The latter is the case found using Ar + H_2 mixture as the principal arc gas.

In one investigation[251)] the nature of the soot formed in the reaction of coal under plasma conditions was examined. X-ray diffraction photographs of the soot showed that some of the lines present in the photograph of the original raw samples were absent, while most of the other lines were reduced in intensity (Fig. 19). Despite the obvious concentration of ash in the soot, the absence or weakening of the lines may be explained by the breaking up of the crystalline structure of the mineral matter and by the transformation of phases that occur at these high temperatures.

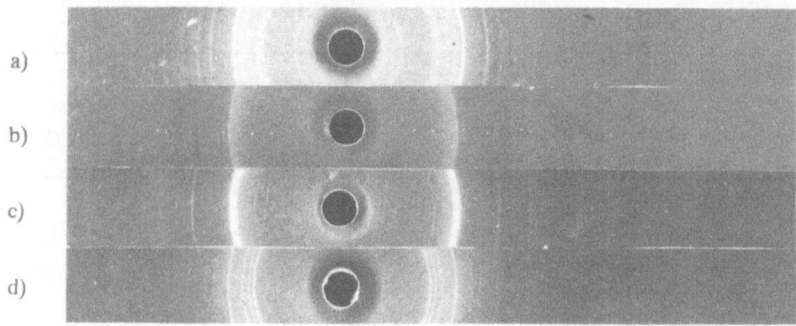

Fig. 19a–d. X-ray powder photographs of
a   original Indian (Jharia) coal,
b   heated at 1500 K,
c   plasma-heated and d) soot. (Photographs courtesy of Dr. S. C. Chakravartty, Central Fuel
Research Institute of India, Dhanbad)

No graphitic character of carbon was detected. The diffraction pattern of the sintered
crust, which was deposited around the anode mouth, showed only the broad bands
of graphite. The degree of graphitization evident from the pattern showed coal particles actually in transition to the high temperature required for efficient conversion
into acetylene. In a subsequent X-ray study[254] the development of considerable
crystalline structure in highly coking coals under instantaneous (15 μs) plasma heating (3000 K) was confirmed.

5.4 Flash and Laser Irradiation of Coal

Coal has been exposed to intense energy bursts using flash and laser techniques. Flash
heating uses an intermittent light source to heat coal particles within 1 ms to temperatures in the range 1300–1800 K. When a high-power laser beam is focussed on
to coal a small inertially confined plasma is generated near the coal surface. Electron
temperatures and densities as high as 5×10^5 K and 1×10^{25} m^{-3}, respectively, have
been measured[255, 256].

5.4.1 Flash Heating of Coal

Several reports[257–264] have been published in which mg samples of pulverized (10μ)
coal placed in fast rotating quartz break-seal tubes were pyrolyzed by exposure to
1 ms high intensity light flash from a xenon-filled quartz helix flash tube capable of
delivering energies up to 4 kJ. After flashing in vacuum or in the presence of added
gases such as Ar, H_2 or N_2 the gaseous products were analyzed by mass spectrometry.
Representative data are given in Table 14. Variations in energy input and particle
size affected the distribution of products which invariably consisted of H_2, CO, CH_4
and C_2H_2. Similar data were also found using the radiant flux from an arc-image fur-

Table 14. Gaseous products from flash irradiation of coal in vacuum[a]

Coal size μ	Energy input, J	Gas composition, mole %							Ref.
		H_2	CH_4	C_2H_2	C_{2s}	C_{3s}	CO	CO_2	
10	800	63.4	6.1	13.6	7.1	0.9	7.1	0.4	259)
10	3200	67.0	1.3	15.5	4.6	0.1	10.2	0.1	259)
200	4000	69.2	4.1	7.8	0.15	0.4	16.1	0.5	260)
10 mm cube[b]	700	62.0	7.0	7.6	2.8	0.3	14.1	5.7	265)

[a] Coal contained 35–40% volatile matter.   [b] Using an arc-image furnace, $C_{2s} = C_2H_4 + C_2H_6$, $C_{3s} = C_3H_6 + C_3H_8$.

nace[265]. The lower rank coals yielded less H_2 and CH_4. When compared with the vaccum runs, added N_2 (10 torr) had no significant effect. Increasing the pressure of added H_2 from 10 to 160 torr decreased the C_2H_2 yield and increased the CH_4 yield. Light pulses from which UV component was filtered produced lesser amount of gases that were richer in methane[261]. There is one patent which describes the flash heating of coal to prepare gaseous and liquid fuels[263].

In one investigation[264] the free radicals present in the residual char from flash-heating of two coal samples with different volatile material content were studied using ESR. The free radical concentration (3×10^{22} kg^{-1}) was higher for the char from the less volatile sample. For both samples the free radical concentration increased with increase in flash energy to a maximum and then decreased slowly. The maximum corresponded to a heat-treatment temperature of 1300 K which was very much higher than the 800–900 K observed for chars from slow heat treatment[266, 267]. The grad-ual decrease in the signal was in contrast to the results for chars prepared by slower heat treatment[266, 267], where an important feature was a rapid decrease in signal with increasing temperature as the char became electrically conducting. Since in the experiment the Q of the ESR cavity was reported to be fairly constant there was no indication of increased electrical conductivity in the char prepared by flash heating.

5.4.2 Laser Irradiation of Coal

Friedel and coworkers[197, 260, 268–273] have published a number of papers on the laser irradiation of coal. In their experiments a 10 mm cube or about 150 mg of < 200 mesh powdered coal was placed in a break-seal tube. For experiments with powdered coal, the tube was rotated to deposit as much powder as possible on the inner surface of the tube in the direct path of the energy. A ruby crystal or rod with an output of about 1.7 J at 6943 Å was used for most of the experiments. Energy from a 4000 J power supply was discharged through the xenon-filled lamp and the 1 ms light pulses were focused on the coal sample to evolve gases and to produce craters of about 300 μ in diameter. The luminous plume was recorded by high-speed photography[272] and the temperature within the plume was estimated to be some-what greater than 1300 K. Presently there is no published information on the photo-

Table 15. Gaseous products from laser irradiation of coal in vacuum[260]

Energy source and/or treatment	Gas composition, mole%								Estimated conversion, wt% coal
	H_2	CH_4	C_2H_2	C_{2s}	C_{3s}	CO	CO_2	HCN	
Ruby laser (1.7 J output)	45.4	4.8	20.9	5.5	5.6	14.4	2.0	1.4	60
Xenon lamp (4 kJ input)	69.2	4.1	7.8	0.2	0.4	16.1	0.5	1.7	18
Carbonization at 1200 K	55.6	31.5	0.1	4.6	0.5	7.4	0.4	–	15

chemical influence of laser energy because duplicate craters have not been produced by different lasers and the temperature effect is very much greater than the light effect. After fractionation the evolved gases were analyzed by mass spectrometry.

Table 15 compares the results of laser irradiation of coal with flash irradiation and conventional carbonization at 1200 K. Laser irradiation yielded higher conversions of coal and greater percentages of acetylene and higher hydrocarbons than flash irradiation or carbonization. When five bursts of 10 J energy were used to irradiate a cube of coal the acetylene yield increased from 20.9 to 25.9 mole %; this constituted 90% of total hydrocarbon products. The decrease in H_2 (Table 15), relative to the product from the flash irradiation, is possibly related to the increase in partially saturated structures such as ethylene (4.9%) and propylene (0.7%). Further, components with molecular weights up to 130 were found in the gas from laser irradiation; diacetylene and vinylacetylene recognized as pyrolysis products of acetylene accounted for 2.4% of the product.

In another series of experiments Friedel and coworkers[268–270] studied the distribution of gaseous products from the laser irradiation of coals of various ranks and particle sizes in various atmospheres. In vacuum experiments the total gas yield varied inversely with coal rank, showing a four-fold increase between anthracite and lignite. The product gas composition as a function of volatile matter in coal is shown in Fig. 20. Yields of acetylene and hydrogen generally increased between anthracite

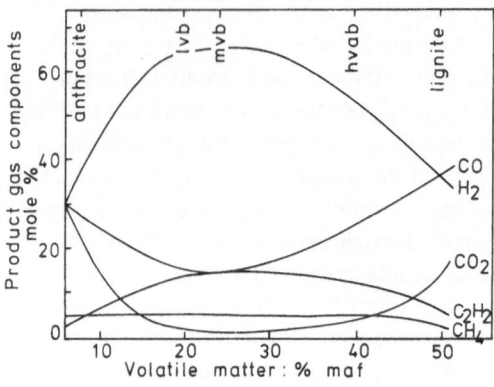

Fig. 20. Distribution of H_2, CO, CO_2, CH_4 and C_2H_2 in the products of laser irradiation of coal as a function of volatile matter in coal. (Redrawn from Karn, F. S., Friedel, R. A., Sharkey, Jr., A. G.: Carbon 5, 25 (1967), by permission of the publishers, Pergamon Press Ltd.)

Table 16. Gaseous products from laser irradiation of coal in various atmospheres (p = 20 torr)[269]

Atmosphere	Gas composition, wt %								
	H_2	CH_4	C_2H_2	C_2H_6	C_{3s}	C_4H_2	CO	CO_2	HCN
Vaccum	6.0	8.5	25.9	3.0	4.1	2.7	41.2	3.7	4.9
He	2.6	7.1	6.7	2.4	21.5	0.5	15.6	43.4	0.2
Ar	4.3	3.1	36.7	0.8	1.1	11.1	26.3	14.3	2.3
H_2	8.3[a]	5.9	25.5	1.0	1.5	0.4	37.2	17.8	2.4
N_2	3.9	4.9	32.6	0.7	0.9	1.6	33.8	19.3	2.3
H_2O vapor	6.1	6.1	22.0	1.1	7.4	1.0	34.6	18.0	3.7

[a] H_2 produced by laser pyrolysis.

and lvb coal, and decreased between hvab coal and lignite. Yields of methane changed relatively little with coal rank. For the smaller particles (10–120 μ) there was a modest increase in methane and an increase in acetylene yield. Table 16 compares the composition of gaseous products from laser irradiation of a hvab coal in vacuum and various atmospheres. The acetylene content of the gas was increased in presence of Ar and N_2, and diminished in presence of He. When the gas pressure was increased the acetylene content increased to 50 wt % with Ar and 45 wt % with N_2, both in the pressure range of 80–400 torr. The role of Ar or N_2 in enhancing the acetylene production is not understood.

In a subsequent investigation[270, 271] the C_2H_2/CH_4 ratio was shown to be related directly to the light flux at the surface of the decomposing coal. Temperatures were changed by varying the energy of the laser beam, the laser focus, and the type of laser. Besides the ruby laser (6943 Å), a neodymium laser (10,600 Å) and a carbon dioxide laser (106,000 Å) were used. Data from the irradiation of coal with these three lasers, including several variations in the energy intensities of the ruby laser, were compared at approximately the same total energy output to determine if there are differences in the quantity and distribution of product gas. The composition of the product gas as a function of light flux is shown in Fig. 21. The temperatures shown in the figure were estimated using gas analyses and assuming the gases to be in equilibrium during their generation. Only the data from irradiations with the CO_2 laser

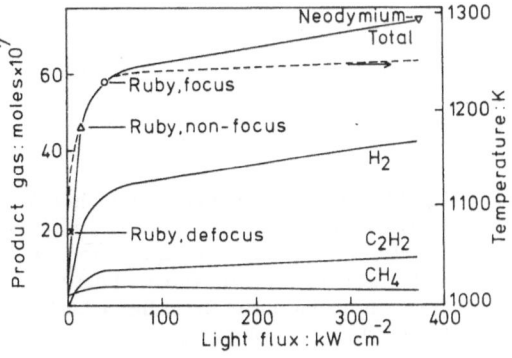

Fig. 21. Total product gas and distribution of H_2, CH_4 and C_2H_2 in the products of laser irradiation of coal as a function of light flux. The broken curve indicates temperatures of laser irradiated coal as estimated from gas analysis. (Redrawn from Karn, F. S., Friedel, R. A., Sharkey, Jr., A. G.: Fuel 48, 297 (1969), by permission of the publishers, IPC Business Press Ltd.)

were inconsistent with the other data, perhaps because of the slow heating and cooling rates arising from the low intensity and rate of emission from the CO_2 laser. The CO_2 laser produced methane in significant amounts but did not produce acetylene which apparently required a higher temperature than was available. The more intense laser beams produced greater quantities of product gases and higher C_2H_2/CH_4 ratios.

The solid residue (53 wt %) from high energy (55 kW/cm^2) ruby laser irradiation of coal analyzed for a C/H atomic ratio of 1.43 which was higher than the value of 1.23 for the unexposed coal[271]. Although the solid did not show most of the IR bands characteristic of the parent coal, its high resolution mass spectrum was similar to that of coal. On the contrary, the solid residue (71 wt %) from low energy (140 W/cm^2) CO_2 laser irradiation of coal was a low-ash, low-density and high-H content solid having a C/H atomic ratio of 1.11. A high resolution mass spectrum showed that it was a complex mixture of a variety of organic compounds, including naphthalenes, phenanthrenes, pyrenes and their methyl-substituted homologues. Compared with conventional carbonization data[271] the solids contained a higher concentration of hydroaromatic compounds.

When the concentrates of macerals of a high-volatile bituminous coal were irradiated with 6-J pulses from a ruby laser the total gas yield varied directly with volatile matter (13.4–55.4 maf %) of the macerals[273]. Major gases evolved were H_2, CO and C_2H_2; their relative concentrations varied little among the macerals.

Laser heating of pulverized coal has been carried out in the source of a time-of-flight mass spectrometer[274, 275]. The pyrolysis products from the coal travelled only about 10 mm in the high vacuum of the mass spectrometer before reaching the ionizing electron beam, thus minimizing the chances of secondary reactions occurring before analysis. At 70 eV the spectra were recorded for approximately 10 ms, starting with a fairly intense spectrum, which increased in intensity during the first ms and then gradually decreased. A typical spectrum of the initial products obtained by heating a hvab coal at two laser energies is shown in Fig. 22 together with probable assignments of the prominent peaks. Peaks with m/e > 100 were not present whether the coal was heated at low or high energies. At high laser energy, only the decomposition products from coal included ions such as Li$^+$, Na$^+$ and K$^+$, radicals such as H· and CH$_3$· and polyacetylenes such as C_4H_2, C_6H_2 and C_8H_2.

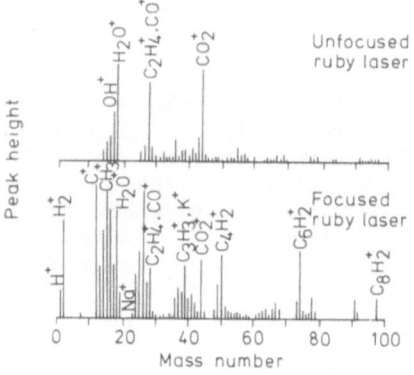

Fig. 22. Mass spectra of products from heating lvb coal with ruby laser at low energy (*unfocused*) and high energy (*focused*). (Redrawn from Joy, W. K., Ladner, W. R., Pritchard, E.: Fuel *49*, 26 (1970), by permission of Dr. Ladner and the publishers, IPC Business Press Ltd.)

5.5 Plasma Gasification of Coal

Interest in coal gasification stems from anticipated serious shortages of natural gas and other forms of energy. The term gasification, as it is presently used, refers not only to heating the coal, as in distillation, but also to the subsequent reaction of the solid residue with air, oxygen, steam or various mixtures. A few papers have been published on the problems of coal gasification in plasma chemical flow reactors[276, 278]. These included the use of electrical coronas[198, 279], electric sparks[280] and plasma jets[250, 281–283]. In some instances[284, 285] plasma-heating was used together with conventional burner heating to improve the gasification.

The distinguishing characteristic of non-plasma and plasma methods of gasification is that the products are different: the non-plasma conventional carbonization process yields gas and tar; plasma processes yield gas and soot. Based on the representative data collated for hvab coal in Table 17 it is inferred that the gas from carbonization is richer in methane, the chief component of natural gas and the preferred product; the gas from plasma is richer in acetylene. In most of the plasma methods collated in Table 17 the gas temperatures were 1300 K or more and any methane formed would readily decompose to yield acetylene. Clearly, in-so-far as the plasma formation of methane is concerned the temperature region of 700–1000 K needs to be investigated. Her 60 Hz discharges, with or without electrodes, in reactors heated to the desired temperature may be advantageous.

With respect to the different coal ranks the only systematic data on gasification which is available at present are for microwave discharges. This data given in Table 18 show that appreciable conversions to gaseous hydrocarbons are found only with hvab

Table 17. Comparison of gas composition for plasma and non-plasma methods of treating hvab coal

Method	Gas composition, mole %				Total gases, wt % coal	Ref.
	H_2	CH_4	C_2H_2	CO		
Carbonization 720 K	19.2	47.8	–	16.0	5	231)
Carbonization 1170 K	55.6	31.5	0.05	7.4	15	260)
Flash irradiation with 4 kJ Xe lamp	69.2	4.1	7.8	16.1	18	260)
Laser irradiation with 1.7 J ruby laser	45.4	4.8	20.9	14.4	60	260)
Arc image furnace	62.0	7.0	7.6	14.1	–	265)
Plasma jet in Ar	48.0	0.8	30.2[a, b]	18.4	20	241, 242)
Microwave discharge in vacuum	64.0	2.2	9.3	22.2	18	231)
in argon	64.0	1.6	10.3	20.7	17	231)
in argon with cooling	12.8	0.3	61.7	4.9	32	231)

[a] Increases to 40% on adding 10% H_2.
[b] By axial injection of coal to 37% in Ar and to 74% in 90% Ar + 10% H_2 (Ref.253)).

Table 18. Extent of gasification of coal vs carbon content[230, 231]

Coal rank	%C	%C in coal							
		Gasified				Converted to gaseous hydrocarbons			
		Vac	Ar[b]	Ar + H$_2$[c]	Ar + H$_2$O[d]	Vac	Ar[b]	Ar + H$_2$[c]	Ar + H$_2$O[d]
Lignite	66.5	20.6	19.4	16.7	20.9	3.9	4.5	4.1	4.1
Lvb	89.6	4.2	4.9	6.7	7.6	2.7	3.5	5.6	4.0
Hvab[a]	81.8	11.1	11.4	13.6	14.8	5.7	6.6	10.4	7.9
		21.7	30.4	24.4	–	21.1	29.3	22.0	–
Anthracite	91.1	1.2	1.2	3.9	8.7	0.6	0.7	3.4	3.0

a Second row obtained with liquid nitrogen cooling.
b p_{Ar} = 5 torr.
c p_{Ar} = 2.2 torr, p_{H_2} = 22 torr.
d p_{Ar} = 8 torr, p_{H_2O} = 10 torr.

coals in Ar + H$_2$ plasmas. With plasma jets it is known[240–243] that carbon conversion to acetylene increases with increasing volatile matter in the coal, that is, with decreasing rank of coal. Whether the major hydrocarbon product is methane or acetylene, other significant products in the gasification are CO and H$_2$. In the thermal gasification processes under development[286, 287] the H$_2$ + CO mixture in the ratio 3:1 is subjected to catalytic methanation. The reaction of carbon monoxide with hydrogen in electrical discharges is of considerable interest and has been studied in some detail. In a static reactor[288], at initial pressures of 12 and 50 torr and with reaction times of the order of a minute, as much as 90% of the CO (even CO$_2$) originally present was converted to hydrocarbons CH$_4$ + C$_2$H$_2$ using microwave discharges in 5 H$_2$: CO mixtures. Even 50 Hz discharges[289] and rf (2–110 MHz) discharges[290] yielded high conversions of H$_2$ + CO to CH$_4$.

Invariably acetylene is the major plasma hydrocarbon product. There is abundant literature on the effect of heat and catalysts on acetylene. Vigorous discharges brought about by rf and microwaves convert acetylene to cuprene[291]. In discharge reaction of H atoms with acetylene a series of reactions occur which consume H atoms and eventually regenerate acetylene[236, 292]. Apparently, then, the acetylene must be alternately hydrogenated and dehydrogenated. Since acetylene is very easily hydrogenated to ethylene by metal catalysts, studies of the catalytic methanation of acetylene may prove interesting.

5.6 Plasma Desulfurization of Coal

The total sulfur content of coal varies in the range of 0.2–11 wt %, although in most cases it is between 1 and 3 wt %[293]. A large part of the sulfur is in the form of FeS$_2$. The amount of organic sulfur is usually one-half to one-third of the total sulfur[294].

All the organic sulfur is believed to be bivalent and it is spread throughout the organic matrix[295]. Very few data on the structure of organic sulfur components and their distribution have been derived by direct observation on coal. Most of the information was obtained from examination of the smaller molecular products which were obtained by breaking the organic coal matrix (depolymerization). Some reviews of the reactions of sulfur in coal-gas reactions can be found in the recent literature[296, 297]. The chemical and physical methods of coal desulfurization was the subject of a recent symposium held by the American Chemical Society[298]. The plasma technique for breaking the organic coal matrix may change the structures of the sulfur constituents as well as those of the hydrocarbon parts.

The two depolymerization methods which have been used most are pyrolysis and hydrogenation. Thiophenes and sulfides were the major sulfur components of tars from coal pyrolysis; hydrogen sulfide and the lower mercaptans and sulfides were found in the volatiles[299]. Hydrogen sulfide and thiophenes were practically the only sulfur products of coal hydrogenation[300]; hydrogen sulfide is produced in char hydrodesulfurization[301].

A number of experiments have been carried out on the synthesis and/or decomposition of hydrogen sulfide in ozonizer discharges[302]. At ordinary temperatures the synthesis was much slower than the decomposition. The higher the ozonizer temperature the lesser was the decomposition, perhaps owing to increased activation of sulfur vapor and the enhanced reverse reaction to form hydrogen sulfide. In presence of NO and CO, hydrogen sulfide was oxidized to sulfur in ozonizer discharges. A method for the simultaneous recovery of H_2 and S from H_2S is described[303]. It used cylindrical stainless steel electrodes in a glow discharge at 825 V and 0.264 A. Very few studies on plasmas in sulfur-containing organic compounds have been made. These are reviewed in Sect. 4.6.

Coffman and Browne[198] reported that in the corona processing of coal, hydrogen sulfide was one of the early products indicating that the sulfur bonds cleaved more readily than any other bond in coal. This is not surprising since Stokes[304] synthesized hydrogen sulfide from its elements in a helium plasma jet and obtained conversions as high as 37% based on the sulfur input.

There are few direct measurements of the sulfur removed from coals treated in plasmas. Zavitsanos and Bleiler[305] reported that coal could be desulfurized without loss of heating value by exposing coal particles of 0.1−4 cm size in 1.5−2 cm layers to 2.45 GHz microwaves for 40 s in air at 1−5 torr pressure. The coal temperature did not exceed 420 K and the sulfur content was reduced from 4.11 to 1.91%. However, Scott et al.[224] did not observe any significant change in S content (2%) of bituminous coals as a result of exposure to 60 Hz discharges in the annular space of a Siemens tube.

The effect of air, steam, hydrogen, carbon monoxide, and ozone on the desulfurization of coals of rank ranging from anthracite to sub-bituminous has been investigated under thermal conditions[306−308]. Air treatment was most effective; the order of desulfurizing ability between 700 and 900 K was air > steam-CO mixture > CO > N_2. At 700 K air removed an average of 38% total sulfur, comprising 51% of the inorganic sulfur and 20% of the organic sulfur. With steam at 900 K, 61% of the total sulfur, 87% of the inorganic and 25% of organic was removed. Hydrogen

was effective at about 1200 K in removing 86% of the total sulfur, 94% of the inorganic and 76% of the organic sulfur. While oxygen and steam are known to oxidize
the pyrite (FeS_2) and the sulfide (FeS) to sulfates and oxides with SO_2 elimination
at temperatures as low as 600–700 K, the reaction of H_2 with FeS_2 becomes important above 800 K only and the reaction of H_2 with FeS becomes significant above
1100 K only. Hydrogenation of thiophene and its derivatives is known to eliminate
H_2S from the molecule[309–311]. Further, thiophenic structures can be formed by
the reactions of sulfur or H_2S with organic molecules or by the reactions of organic
molecules like C_2H_2 and C_2H_4 with FeS_2 [312]. All these reactions need to be investigated under plasma conditions as well.

6 Other Fossil Fuel Plasmas

Fossil fuels which have not been included in the preceding sections are tar or heavy
oil sands, oil shales and Gilsonite. The tar or heavy oil sands are impregnated with
heavy crude oil that is too viscous to permit recovery by natural flow into wells.
Athabasca tar sands[227] contain 12% toluene soluble bitumen which analyzes for
83.2% C, 10.4% H, 0.9% O, 0.4% N and 4.0% S. Oil shales differ from tar sands in
that their hydrocarbon contents, usually aliphatic or alicyclic, are in a solid rather
than a viscous liquid form. From oil shale dposits an intricately polymerized sub-
stance called *kerogen* is prepared[227]. Typical analyses of these solids, on a moisture-
free basis, are: 21.1% C, 2.9% H, 0.7% N, 1.3% S, 14.4% O and 59.6% ash for oil
shale; 66.4% C, 8.8% H, 2.1% N, 3.3% S, 5.2% O and 14.2% ash for kerogen[227, 313].

Table 19. Static pyrolysis in microwave discharge[227, 313]

Product	Tar sands		Oil shale		Kerogen		Gilsonite	
mmol/g solid	I	II	I	II	I	II	I	II
H_2	1.93	2.77	11.6	11.4	20.6	17.2	16.7	5.77
CH_4	0.07	0.12	0.91	0.75	2.09	1.83	1.24	2.63
C_2H_2	0.19	0.48	1.51	3.2	3.54	5.88	2.36	5.56
C_2H_6	0.04	0.02	0.08	0.05	0.29	0.08	0.21	0.12
CO	0.82	0.53	4.97	4.97	2.64	2.71	0.56	0.55
CO_2	0.02	0.07	0.35	0.22	0.12	0.36	0.01	0.04
HCN[a]	0.01	0.03	0.28	0.37	0.72	1.11	0.52	1.21
H_2O	0.02	0.07	0.18	0.12	0.33	0.04	0.16	0.17
Total gases, wt% solid	3.7	3.9	25.9	28.1	31.8	34.8	16.5	25.5

I. Experiments in vacuum; reaction time 10–20 min.
II. Experiments in the presence of 25 torr of H_2 + Ar (4:1) mixture: reaction time 2 min. For
H_2 product the net increase is given.
a Peaks attributed to ethylene were neglected.

Gilsonite is a natural black bitumen found in Utah and Colorado. A typical sample analyzes for 84.7% C, 10.3% H, 3.0% N, 0.19% S and 1.4% O[227, 313]. Only one investigation[227, 313] on the plasma gasification of these fuels has been reported. This included both static pyrolysis and flow reactions in a 2.45 GHz microwave cavity discharge operated at 50 W.

In static experiments, the discharge was initiated with 10 mg sample either in vacuum (10^{-4} torr) or in the presence of Ar (5 torr) and a $4H_2$: Ar mixture (25 torr). In each case the pressure reached a plateau within a few minutes. Oil shale and kerogen which contained far greater amounts of ungasifiable minerals produced significantly greater amounts of gases than hvab coal. In the presence of Ar gas evolution proceeded at a higher rate and reached a limiting value quickly. The mass spectrometric analysis of the products are given in Table 19. Hydrogen and carbon monoxide were the major products, expect for low-oxygen containing Gilsonite. Acetylene and methane were invariably the major hydrocarbon products. The yield of acetylene increased in presence of added H_2 + Ar for all these fuels, but methane yields were increased only for Gilsonite and tar sands. For both oil shales and kerogen the methane yields decreased on addition of H_2 + Ar.

In flow experiments, argon was passed through a reactor (Fig. 9a) containing 100 mg of sample in the discharge zone at a pressure of 10 torr and a flow rate of 1 cm^3/min. The gasification was nearly complete after 30 min, except for slow degassing of very small amounts of hydrogen. As in the static experiments, mass spectrometric analysis showed that H_2 and CO were the major products, methane, acetylene and ethylene being the major hydrocarbon products. Variations in the rate of gas production from oil shale and kerogen for H_2, CO, CH_4 and C_2H_2 + C_2H_4 products are shown in Fig. 23. The other fuels exhibited similar characteristics of gas evolution, the initial high rate of gasification decreasing rapidly with the time of discharge. A comparison of the rate curves for various fossil fuels showed that total moles of gases evolved per unit weight of the solid decreased in the following order: kerogen, Gilsonite, oil shale and hvab coal. Kerogen and Gilsonite evolved mostly hydrogen, but more gases than other fuels perhaps because of their high H contents. As expected small amounts of CO were evolved from kerogen and tar sands because of the low oxygen content of these materials.

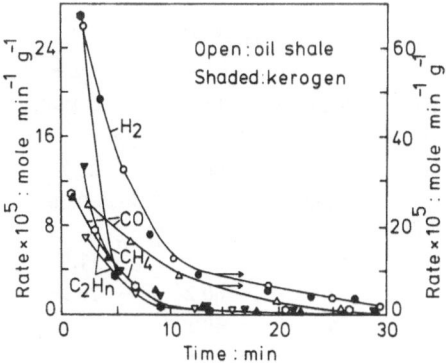

Fig. 23. Rate of gas production from oil shale (*open points*) and kerogen (*shaded points*). ○, H_2; △, CO; ○, CH_4; ▽, C_2H_n (= C_2H_2 + C_2H_4). (Redrawn from Fu, Y. C., Balustein, B. D., Wender, I.: Chem. Eng. Progr. Symp. Ser. *67*, 47 (1971), by permission of the authors and the publishers, the American Institute of Chemical Engineers)

Table 20. Condensable products from microwave discharge-flow systems[227]

Product[a] mmole/g solid	Tar sands	Oil shale	Kerogen	Gilsonite
C_2H_2	0.41	1.07	2.02	2.08
C_2H_4	0.11	1.42	3.09	0.55
C_2H_6	0.04	0.2	0.29	0.38
C_3H_6	0.02	0.25	0.63	0.16
C_3H_8	0.01	0.05	0.13	0.05
C_4H_2	0.03	0.16	0.31	0.12
C_6H_6	0.02	0.05	0.13	0.08
CO_2	0.06	1.85	0.45	0.03
HCN	0.02	0.13	0.45	0.43
H_2O	0.36	0.23	0.71	0.13
Total condensables, wt% solid	2.9	18.8	25.1	11.9

[a] Reaction time of 30 min except 60 min for Gilsonite.

Mass spectrometric analysis of the discharge products condensed in a liquid nitrogen trap showed that these consisted mainly of C_2H_2, C_2H_4, higher molecular weight hydrocarbons (up to C_8, including cyclic compounds), CO_2, HCN and H_2O (Table 20). It is not known whether the higher hydrocarbons were formed by free radical reactions in the cold trap or as a result of the quick removal of product from the discharge zone thus avoiding extensive fragmentation of the larger molecules formed in the plasma. More acetylene was found for kerogen than for other fuels; both oil shale and kerogen yielded more ethylene than acetylene.

Data on the extent of gasification are summarized in Table 21. The non-condensable H_2, CH_4 and CO apparently accounted for the differences between the weight loss of solid and the weight of condensable products. Comparison of the data on weight loss and ash content for each solid indicated that large parts of the organic

Table 21. Degree of gasification of fuels in argon discharge-flow systems[227]

	Hvab coal	Tar sand	Oil shale	Kerogen	Gilsonite
Carbon, %	81.9	9.98	21.1	66.4	84.7
Ash, %	2.1	88	59.6	14.2	0.36
Total condensables, wt% solid	9.5	2.9	18.8	25.1	11.9
Actual wt loss of solid, %	35.5	5.3	34.4	60.2	20.9
Ash-free solid gasified, %	36.2	44.2	85.3	69.1	21.0

matter in oil shale and kerogen were gasified in the discharge. Further, irrespective of the inherent structural differences the extent of gasification of fuels in the discharge decreased with increasing carbon content of the solid. Although volatile organic matter was removed it was difficult to strip the complex structure of coal in low power plasmas even with the use of high frequency. This apparently accounted for the substantial amounts of fixed carbon in coal which remained ungasified.

7 Concluding Remarks

The impression gained from the preceding sections is that insufficient attention paid to experimental parameters may be the reason for the diversity of results obtained in the systems considered. Meaningful comparisons of published results by different workers are difficult because often the magnitude of parameters known to be important is not given, and in some cases at least, not even measured or considered. Further experimental work and theoretical modelling under more carefully controlled conditions are required for comprehensive understanding of the behavior of fossil fuel plasmas.

The status of work on fossil fuel plasmas indicates that some aspects merit detailed study. This will allow one to define better the scope plasma techniques offer for solving fossil energy problems. From the technological point of view line-frequency (60 Hz) operated plasmas are convenient but economically inefficient. High frequency discharges are less desirable economically. The plasma jet appears to be the most promising for acetylene and ethylene production. Optimizing the conditions for ethylene production in natural gas and petroleum plasmas is of considerable interest in view of its industrial importance. Several new experiments should be performed for desulfurization of petroleum fractions in presence of additives which readily combine with sulfur atoms or hydrogen sulfide in plasmas. These experiments could parallel those already done by thermal pyrolysis.

With regard to coal it appears that the temperature range of 700–1000 K must be achieved in plasmas for its gasification predominantly to methane. This intermediate temperature range is achieved neither in conventional low temperature electric discharges nor in high temperature arcs and plasma jets easily. It appears that a combination of heating and electrical discharge is desirable for this purpose. In some laser experiments with coal these temperatures were attained and methane produced. Optimization of the conditions for methane formation from coal and coal combustion products (CO and CO_2) by the plasma technique should be attempted in view of the anticipated short supply of natural gas. Besides gasification the organic chemistry of coal in plasmas is fascinating in its own right. The presence of catalysts in the plasma zone or downstream might produce new reactions and novel chemicals. Finally, experiments on partial rather than total removal of sulfur from coal by plasma technique may provide a solution to the existing environmental problems in coal burning.

8 References

1. Reactions under plasma conditions, Venugopalan, M. (ed.). New York: Wiley-Interscience 1971
2. Rossini, F. D.: Selected values of physical and thermodynamic properties of hydrocarbons and related compounds. Pitsburgh: Carnegie Press 1953; JANAF thermochemical tables, Stull, D. R. (project director). Midland, Michigan: Dow Chemical Company 1965
3. Jones, W. I.: J. Inst. Fuel. *37*, 3 (1964)
4. Plooster, M. N., Reed, T. B.: J. Chem. Phys. *31*, 66 (1959); Duff, R. E., Bauer, S. H.: J. Chem. Phys. *36*, 1754 (1962)
5. Baddour, R. F., Blanchet, J. L.: Ind. Eng. Chem., Process Design Develop. *3*, 258 (1964)
6. Clarke, J. T., Fox, B. R.: J. Chem. Phys. *46*, 827 (1967); Clarke, J. T.: in: The application of plasmas to chemical processing. Baddour, R. F., Timmins, R. S. (ed.). p. 140. Cambridge, MA.: M.I.T. Press 1967
7. Vepřek, S., Haque, M. R.: Appl. Phys. *8*, 303 (1975); Vepřek, S., Haque, M. R., Oswald, H. R.: J. Nucl. Mater. *63*, 405 (1976)
8. Marynowski, C. W., Philips, T. C., Philips, J. R., Hiester, N. K.: Ind. Eng. Chem., Fundamentals *1*, 52 (1962); Griffiths, D. M. L., Standing, H. A.: Adv. Chem. Ser. *55*, 666 (1966)
9. Timmins, R. S., Ammann, P. R.: in: The application of plasmas to chemical processing. Baddour, R. F., Timmins, R. S. (ed.). p. 111. Cambridge, MA.: M. I. T. Press 1967
10. Vepřek, S.: J. Crystal Growth *17*, 101 (1972); Z. physik. Chem. (Neue Folge) *86*, 95 (1973); in: Topics in Current Chemistry, Vol. 56. Berlin – Heidelberg – New York: Springer 1975
11. Encyclopedia of chemical technology, 2nd ed., Kirk, R. E., Othmer, D. F. (ed.). Vol 10, p. 450. New York: Wiley 1966
12. Gas engineer's handbook. p. 2. New York: Industrial Press 1965
13. Fischer, F., Peter, K.: Z. physik. Chem. *A141*, 180 (1929)
14. Brewer, A. K., Kueck.: J. Phys. Chem. *35*, 1293 (1931)
15. Peters, K., Wagner, O. H.: Z. physik. Chem. *A153*, 161 (1931)
16. Yeddanapalli, L. M.: J. Chem. Phys. *10*, 249 (1942)
17. Wiener, H., Burton, M.: J. Amer. Chem. Soc. *75*, 5815 (1953)
18. Tickner, A. W.: Can. J. Chem. *39*, 87 (1961)
19. Drost, H., Klotz, H. D., Timm, U.: Proc. 13th Intern. Conf. Phenom. Ioniz. Gases *1*, 85 (1977)
20. Kraaijveld, H. J., Waterman, H. I.: Brennstoff-Chem. *42*, 369 (1961)
21. Schmellenmeier, H., Roth, L., Schirrwitz, H., Wolff, J.: Chem. Tech. (Berlin) *15*, 580 (1963); *16*, 33 (1964)
22. Amouroux, J.: Vide *26* (154), 164 (1971)
23. Vishnevetskii, I. I., Semkin, B. V.: Tezisy. Dokl.-Vses. Simp. Plazmokhim., 2nd, *1*, 108 (1975)
24. Bykova, L. A., Vishnevetskii, I. I., Semkin, B. V., Kravtsov, A. V., Grunin, V. K.: Tezisy. Dokl.-Vses. Simp. Plazmokhim., 2nd, *2*, 140 (1975)
25. Smol'yaninov, S. I., Kravtsov, A. V., Vishnevetskii, I. I., Kotlova, L. F.: Izv. Tomsk. Politekhn. In-fa (253), 70 (1976)
26. Ponnamperuma, C., Woeller, F.: Nature *203*, 272 (1964)
27. Ponnamperuma, C., Pering, K.: Nature *209*, 979 (1966)
28. Ponnamperuma, C., Woeller, F., Flores, J., Romiez, M., Allen, W.: Adv. Chem. Ser. *80*, 280 (1969)
29. Friedmann, N., Miller, S. L.: Science *166*, 766 (1969)
30. Friedmann, N., Bovee, H. H., Miller, S. L.: J. Org. Chem. *36*, 2894 (1971)
31. Vishnevetskii, I. I., Bykova, L. A., Kotlova, L. F., Semkin, B. V., Smol'yaninov, S. I.: Izv. Tomsk. Politekhn. In-fa (300), 91 (1977)
32. Tsentsiper, A. B., Eremin, E. N., Kobozev, N. I.: Dokl. Akad. Nauk SSSR *141*, 117, 378 (1961)
33. Borisova, E. N., Eremin, E. N.: Zh. Fiz. Khim. *36*, 1261 (1962)
34. Tsentsiper, A. B., Eremin, E. N., Kobozev, N. I.: Zh. Fiz. Khim. *37*, 835 (1963)

35. Meskova, G. I., Eremin, E. N.: Khim. Fiz. Nizkotemp. Plasmy, Tr. Mezhvuz. Konf., 1st, 202 (1970)
36. Gartaganis, P. A., Winkler, C. A.: Can. J. Chem. *34*, 1457 (1956)
37. Shekhter, A. B.: Chemical reactions in electrical discharge. Moscow: Dept. Sci. Techn. Infn. 1935
38. Karpukhin, P. P., Slominskii, L. I.: Tr. Khar'kovsk. Politekhn. Inst. *39*, 5 (1962)
39. De Paoli, S., Strausz, O. P.: Can. J. Chem. *48*, 3756 (1970)
40. Alcock, W. G., Hayward, E. J., Mile, B., Ward, B.: Can. J. Chem. *50*, 3813 (1972)
41. Eremin, E. N.: Zh. Fiz. Khim. *32*, 2543 (1958)
42. McCarthy, R. L.: J. Chem. Phys. *22*, 1360 (1954)
43. Vastola, F. J., Wightman, J. P.: J. Appl. Chem. *14*, 69 (1964)
44. Wightman, J. P., Johnston, N. J.: Adv. Chem. Ser. *80*, 322 (1969)
45. Baddour, R. F., Dundas, P. H.: in: The application of plasmas to chemical processing. Baddour, R. F., Timmins, R. S. (ed.). p. 93. Cambridge, MA.: M. I. T. Press 1967
46. Miquel, R., Chirol, M.: Bull. Soc. Chim. France, 1677 (1962)
47. LeGoff, F., Vuillermoz, B.: J. Chim. Phys. Physicochim. Biol. *66*, 403 (1969)
48. Simionescu, Cr. I., Dumitriu, S., Bulacovschi, V., Onac, D.: Z. Naturforsch. *30b*, 516 (1975)
49. Popovici, C.: Proc. Symp. Electron Vacuum Phys., Balatonfoldvar, Hungary, 451 (1962)
50. Simionescu, Cr. I., Dumitriu, S., Popa, V. I.: Z. Naturforsch. *31b*, 466 (1976)
51. Studniarz, S. A., Franklin, J. L.: J. Chem. Phys. *49*, 2652 (1968)
52. Smolinsky, G., Vasile, M. J.: Int. J. Mass Spectrom. Ion Phys. *16*, 137 (1975)
53. Vasile, M. J., Smolinsky, G.: Int. J. Mass Spectrom. Ion Phys. *18*, 179 (1975)
54. Starodubtsev, S. V., Ablyaev, Sh. A., Keitlin, L. G.: Izv. Akad. Nauk Uz.SSR, Ser. Fiz.-Mat. Nauk *6* (5), 50 (1962)
55. Starodubtsev, S. V., Ablyaev, Sh. A., Bakhramov, F., Zipatdinov, Sh., Keitlin, L. G.: Izv. Akad. Nauk Uz.SSR, Ser. Fiz.-Mat. Nauk *6* (5), 58 (1962)
56. Sheer, C.: in: Vistas in science. Arm, D. L. (ed.). p. 135. Albuquerque: University of New Mexico Press 1968
57. Sheer, C., Korman, S.: Adv. Chem. Ser. *131*, 42 (1974)
58. Leutner, H. W., Stokes, C. S.: Ind. Eng. Chem. *53*, 34, 341 (1961)
59. Anderson, J. E., Case, L. K.: Ind. Eng. Chem., Process Design Develop. *1*, 161 (1962)
60. American Cyanamid Co.: French Pat. 1,323,474 (April 5, 1963)
61. Kokurin, A. D., Obezkov, V. D., Kolodin, E. A.: USSR Pat. 162,114 (April 16, 1964)
62. Chemische Werke Huels, A.-G.: French Pat. 1,396,736 (April 23, 1965)
63. Pollock, L. W., Begley, J. W.: U.S. Pat. 3,248,446 c April 26, 1966)
64. Colton, J. W.: U.S. Pat. 3,256,358 (June 14, 1966)
65. AGA & AB: French Pat. 2,202,722 (May 10, 1974)
66. Fey, M. G., Kemeny, G. A., Azinger, F. A., Jr.: German Pat. 2,634,616 (February 24, 1977)
67. Kawana, Y.: Chem. Econ. Eng. Rev. *4*, 13 (1972)
68. Eremin, E. N., Kobozev, N. I., Lyudkosvskaya, B. G.: Zh. Fiz. Khim. *32*, 2315, 2767 (1958)
69. Eremin, E. N.: Khim. Prom., 73 (1958)
70. Tsentsiper, A. B., Eremin, E. N., Kobozev, N. I.: Zh. Fiz. Khim. *37*, 1487 (1963)
71. Il'in, D. T., Eremin, E. N.: Zh. Prikl. Khim. *35*, 2064, 2496 (1962); *38*, 2479, 2774 (1965)
72. Il'in, D. T., Sidorov, V. I., Uryukov, B. A., Fridberg, A. E.: Zh. Prikl. Khim. *42*, 648 (1969)
73. Il'in, D. T., Sidorov, V. I., Uryukov, B. A., Fridberg, A. E.: Khim. Fiz. Nizkotemp. Plazmy, Tr. Meshvuz. Konf., 1st, 282 (1970)
74. Suris, A. L., Shorin, S. N.: Khim. Vys. Energ. *1*, 264 (1967)
75. Agaeva, F. M., Kerimov, A. M.: Tr. Azerb. Nauch-Issled. Inst. Energ. (19), 151 (1970)
76. Nakashio, F., Takahashi, T.: Kagaku Kogaku *36*, 774 (1972)
77. Kerimov, A. M., Mirzoeva, L. M., Musolin, V. N.: Izv. Vyssh. Uchleb. Zaved. Neft. Gaz. *17* (4), 68 (1974)
78. Okabayasi, T., Naito, P., Tozaki, Y., Kato, Y.: Kogyo Kagaku Zasshi *74*, 2057 (1971)
79. Kobozev, Yu. N., Khudyakov, G. N.: Gazov. Prom. *16* (2), 40 (1971)
80. Kinetika i termodinam. khim. reaktsii v nizkotemperaturnoi plazme, Polak, L. (ed.). Moscow: Nauka 1965

81. Vursel, F., Polak, L.: in: Reactions under plasma conditions, Vol. II. Venugopalan, M. (ed.). p. 299. New York: Wiley-Interscience 1971
82. Gulyaev, G. V., Kozlov, G. I., Polak, L. S., Khitrin, L. N., Khudyakov, G. N.: Neftekhimiya 2, 793 (1962)
83. Polak, L. S., Neftekhimiya 7, 463 (1967)
84. Polak, L. S., Gulyaev, G. V.: Vestn. Akad. Nauk SSSR 38, 39 (1968)
85. Polak, L. S.: Khim. Atsetilena, Tr. Vses. Konf., 3rd, 362 (1968)
86. Valibekov, Yu. V., Vursel, F. B., Gutman, B. E., Polak, L. S.: Khim. Tadzh., 68 (1973); Khim. Vys. Energ. 7, 211 (1973)
87. Nursultanov, O. S., Polak, L. S., Popov, V. T.: Tezisy Dokl.-Vses. Simp. Plazmokhim., 2nd, 2, 205 (1975)
88. Vursel, F. B., Polak, L. S., Epshtein, I. L.: Khim. Vys. Energ. 10, 234 (1976)
89. Szymanski, A., Lisicki, Z., Minc, S.: Koks, Smola, Gaz 10, 320 (1965)
90. Minc, S., Szymanski, A., Varykha, S.: U.S. Clearinghouse Fed. Sci. Tech. Inform. AD653539 (1967)
91. Ryszka, E., Sobonski, M.: Hutnik 38, 616 (1971)
92. Andreev, D. A., Bykov, Yu. M., Dyukov, V. G., Likhter, A. M.: Khim. Mashinostr. 4, 168 (1975)
93. Ganz, S. N., Doltoratski, M. N., Volodin, I. S., Parkhomenko, V. D.: Khim. Tekhnol. (Kharkov), (23), 17 (1971)
94. Zubkova, K. A., Laktyushin, A. N., Yas'ko, O. I.: Tezisy Dokl.-Vses. Simp. Plazmokhim., 2nd, 2, 172 (1975)
95. Gershuni, S. Sh., Suris, A. L., Shorin, S. N.: Khim. Vys. Energ. 9, 528 (1975)
96. Szymanski, A.: Bull. Acad. Pol. Sci. Ser. Sci. Chim. 26, 349 (1978)
97. Yamamoto, T.: Yûki Gôsei Kagaku Kŷokaishi 14, 461 (1956)
98. Dolal, V.: Khim. Tverd. Topl. 3, 97 (1978)
99. Maksimov, E. V., Suleimenov, A. B., Mikhailov, V. V., Plitsyn, V. T.: in: Vosstanov.-Tepl. Obrab. Zhelezorudn. Margantsebogo Syr'ya. Agreev, N. V. (ed.). p. 123. Moscow: Nauka 1974
100. Leutner, H. W.: Ind. Eng. Chem. Process Design Develop. 2, 315 (1963)
101. Stokes, C. S., Correa, J. J., Streng, L. A., Leutner, H. W.: A.I.Ch.E. Journal 11, 370 (1965)
102. Freeman, M. P., Skrivan, J. F.: A.I.Ch.E. Journal 8, 450 (1962)
103. Hulburt, H. M., Freeman, M. P.: Trans. N. Y. Acad. Sci. 2 (25), 770 (1963)
104. Freeman, M. P.: Adv. Chem. Ser. 80, 406 (1969)
105. Hoffmann: Proc. Int. Conf. Phenom. Ionized Gases, 13th, 1, 217 (1977)
106. Bronfin, B. R.: Adv. Chem. Ser. 80, 423 (1969)
107. Nishimura, Y., Takeshita, K., Sakai, W.: Sekuju Gokkai Shi 15, 299 (1972)
108. Valibekov, Yu. V.: Dokl. Akad. Nauk Tadzh. SSR 8 (10), 25 (1965)
109. Valibekov, Yu. V., Bolotov, G. M.: Izv. Akad. Nauk Tadzh. SSR, Otd. Fiz.-Mat. Geol. Khim. Nauk, (2), 47 (1968)
110. Valibekov, Yu. V., Bolotov, G. M.: Gazov, Prom. 14 (3), 38 (1969)
111. Valibekov, Yu. V., Bolotov, G. M.: Dokl. Akad. Nauk Tadzh. SSR, 13 (3), 27 (1970)
112. Valibekov, Yu. V., Gutman, B. E., Chuprina, G. A., Tursunov, B. V., Mirzabaev, G. A.: Dokl. Vses. Konf. Khim. Atsetilena, 4th, 3, 403 (1972)
113. Valibekov, Yu. V., Gutman, B. E., Tursunov, B. V., Chuprina, G. A.: Dokl. Akad. Nauk Tadzh. SSR 15 (3), 32 (1972)
114. Valibekov, Yu. V., Gutman, B. E., Polak, L. S.: Khim. Vys. Energ. 7, 60 (1973)
115. Valibekov, Yu. V., Gutman, B. E., Chuprina, G. A.: Izv. Akad. Nauk Tadzh. SSR, Otd. Fiz.-Mat. Geol.-Khim. Nauk, (2), 45 (1973)
116. Valibekov, Yu. V., Gutman, B. E.: Khim. Tekhnol. Topl. Masel, (7), 18 (1975)
117. Valibekov, Yu. V., Gutman, B. E., Priimak, A. D., Kononov, V. V., Novikova, G. A.: Izv. Akad. Nauk Tadzh. SSR, Otd. Fiz.-Mat. Geol. Khim. Nauk, (4), 95 (1976)
118. Valibekov, Yu. V., Priimak, A. D., Novikova, G. A., Kononov, V. V.: Izv. Akad. Nauk Tadzh. SSR, Otd. Fiz.-Mat. Geol.-Khim. Nauk, (3), 129 (1977)

119. Gorokhovskii, A. V., Lipovskii, I. M., Sverdlov, L. M.: Izv. Vyssh. Uchebn. Zaved Khim. Khim. Tekhnol. *21* (5), 652 (1978)
120. Venugopalan, M.: Chemistry and our world. p. 211. New York: Harper and Row 1975
121. Rossini, F. D.: J. Chem. Ed. *37*, 558 (1960)
122. Linder, E. G.: Phys. Rev. *36*, 1375 (1930)
123. Harkins, W. D., Jackson, J. M.: J. Chem. Phys. *1*, 37 (1933)
124. Cross, P. E.: Nature *208*, 892 (1965)
125. Brooks, J. D., Hesp, W. R.: Australian Chem. Process Eng. *21* (2), 16 (1968)
126. Venugopalan, M., Scott, T. W.: Z. Physik. Chem. (Neue Folge) *108*, 157 (1977)
127. Kleyer, D. L., Venugopalan, M.: Unpublished
128. Rowland, H. R.: Elec. Eng. *50*, 288 (1931)
129. Coates, A. D.: U.S. Dept. Com., Office Tech. Serv. AD419-618 (1962)
130. Vastola, F. J., Wightman, J. P.: J. Appl. Chem. *14*, 69 (1964)
131. Venugopalan, M., Hsu, P. Y.-W.: Z. Physik. Chem. (Neue Folge) *102*, 127 (1976)
132. Williams, T., Hays, M. W.: Nature *209*, 769 (1966)
133. Jesch, K., Bloor, J. E., Kronick, P. L.: J. Polymer Sci. *A14*, 1487 (1966)
134. Il'in, D. T., Eremin, E. N.: Vestn. Mosk. Univ. Ser. II, Khim. *17* (2), 29 (1962); Zh. Prikl. Khim. *38*, 2786, 2876 (1965)
135. Schmellenmeier, H., Roth, L., Schirrwitz, H., Wolff, J.: Chem. Tech. *15* (11), 659 (1963)
136. Polak, L. S.: Proc. of World Petroleum Congr., 7th, 283 (1967); 8th, 367 (1971)
137. Polak, L. S., Endjuskin, P. N., Uglev, V. N., Volodin, N. L.: Khim. Vys. Energ. *3*, 184 (1969)
138. Vursel, F. B., Polak, L. S., Epshtein, I. L.: Tezisy Dokl.-Vses. Simp. Plazmokhim., 2nd 156 (1975)
139. Rozanova, M. V., Sidorov, V. I.: Khim. Prom-St. (Moscow), (2), 100 (1976)
140. Valibekov, Yu. V., Novikova, G. A., Knasova, E. I., Priimak, A. D.: Izv. Akad. Nauk Tadzh. SSR, Otd. Fiz.-Mat. Geol.-Khim. Nauk, (3), 129 (1977)
141. Gehrmann, K., Schmidt, H.: Proc. of World Petroleum Congr., 8th, 379 (1971)
142. Fedoseev, S. D., Shteiner, E. V.: Tr. Mosk. Khim.-Tekhnol. Inst. *48*, 172 (1965)
143. Plasma-Fax Bulletin PF-2, Thermal Dynamics Corp., Lebanon, N. H. (October 1960)
144. Krylova, I. A., Moss, A. L., Shcherbina, E. T., Efimova, T. I.: Khim. Vys. Energ. *6* (2), 157 (1972)
145. Zabrodin, V. K., Krylova, I. A., Moss, A. L., Ermolaeva, I. M.: Khim. Vys. Energ. *7* (6), 491 (1973)
146. Pohl, D., Koehler, R., Koch, J.: East German Pat. 2,257,583 (June 28, 1973)
147. Horowitz, M.: U.S. Pat. 3,902,071 (August 26, 1975)
148. Ishibashi, W., Moriyama, S.: Japanese Pat. 7,756,078 (May 9, 1977)
149. Pilz, K.: German Pat. 2,639,807 (March 9, 1978)
150. Khudyakov, G. N., Ykovlev, G. G., Pendrakovskii, V. T., Kobozev, Yu. N., Lytkin, A. Yu.: U.S.S.R. Pat. 439,142 (February 5, 1978)
151. Fey, M. G.: German Pat. 2,731,042 (January 19, 1978)
152. Pechuro, N. S., Merkur'ev, A. N.: Probl. Elektr. Obrabotki Materialov, Akad. Nauk SSSR, Tsentr. Nauchn. Issled. Lab. Elektr. Obrabotki Materialov, 199, 214 (1962)
153. Pechuro, N. S., Grodzinskii, E. Ya., Pesin, O. Yu.: Probl. Elektr. Obrabotki Materialov, Akad. Nauk SSSR, Tsentr. Nauchn. Issled. Lab. Elektr. Obrabotki Materialov, 209 (1962)
154. Pechuro, N. S., Merkur'ev, A. N., Grishin, A. N.: Sintez i Svoislva Monomerov, Akad. Nauk SSSR, Inst. Neftekhim. Sinteza, Konf. po. Vysokomolekul, Soedin, 22 (1962); Elektroiskovaya Obrabotka Metal., Akad. Nauk SSSR, Tsent. Nauchn-Issled. Lab. Elektr. Obrabotki Metal, 88 (1963)
155. Pechuro, N. S., Grodizinskii, E. Ya., Pesin, O. Yu., Estrin, R. I.: U.S.S.R. Pat. 172,766 (July 7, 1965)
156. Pechuro, N. S., Soldatenkov, A. T., Merkur'ev, A. N.: Elektron. Obrab. Mater., (2), 41 (1969)
157. Pechuro, N. S., Pesin, O. Yu., Filimonov, V. A.: Khim. Reakts. Org. Prod. Elek. Razryadakh. Akad. Nauk SSSR, Eksp. Nauch.-Issled. Inst. Metallorezhushchukh Stankov, 101 (1966)
158. Pechuro, N. S., Pesin, O. Yu., Ushakov, A. L.: Gazov. Prom. *12* (4), 39 (1967)

159. Pechuro, N. S., Pesin, O. Yu.: U. S. Pat. 3,519,551 (July 7, 1970)
160. Pechuro, N. S., Pesin, O. Yu., Khachatryan, Ts. A., Gzararyan, G. P., Estrin, R. I.: Prom Arm (4), 15 (1972)
161. Pechuro, N. S., Pesin, O. Yu., Khachatryan, Ts. A., Gzararyan, G. P.: Prom Arm (4), 17 (1973)
162. Pechuro, N. S., Pesin, O. Yu., Khachatryan, Ts. A., Gzararyan, G. P.: Tezisy Dokl.-Vses. Konf. Khim. Atsetilena, 5th, 479 (1975)
163. Von Edinger, W.: U.S. Pat. 2,632,731 (March 24, 1953)
164. Stunemann, E., Lanz, T. E.: U.S. Pat. 2,879,218 (March 24, 1959)
165. Wangelin, D. J., Bowen, E. C.: U.S. Pat. 2,900,321 (August 18, 1959)
166. Ishibashi, W.: Japanese Pat. 6,821,282 (Sept. 12, 1968)
167. Lonza Elektrizitätswerke u. Chemische Fabriken, A.G.: Brit. Pat. 783,819 (October 2,1957)
168. Deutsche Gold u. Silber-Scheideanstalt vorm Roessler: German Pat. 960,057 (March 28, 1957); Brit. Pat. 794,397 (April 30, 1958), 805,608 (December 10, 1958)
169. Sulzer Brothers Ltd.: Swiss Pat. 317,876 (January 31, 1957)
170. Chemische Werke Huels A.G.: Brit. Pat. 792,604 (April 2, 1958), 804,090 (November 5, 1958)
171. Twatani and Co., Ltd.: Belg. Pat. 665,833 (October 18, 1965), French Pat. 1,444,373 July 1, 1966)
172. Inoue, K.: Japanese Pat. 7,445,682 (December 5, 1974)
173. Heinrich Koppers GmbH: German Pat. 1,158,046 (November 28, 1963)
174. Firma Hans J. Zimmer Verfahrenstechnik: French Pat. 1,323,125 (April 5, 1963)
175. Yanagisawa, E.: Japanese Pat. 7,441,401 (April 18, 1974), 7,801,242 (January 17, 1978)
176. Andrussow, L.: Erdoel u. Kohle *12*, 24 (1959)
177. Kokurin, A. D.: Khim. Prom. *42*, 725 (1966)
178. Trufanov, P. A., Kokurin, A. D., Kolodin, E. A.: Tr. Nauch-Issled. Inst. Slanstev (17), 99 (1968)
179. Furuta, T., Sanada, Y., Honda, H.: Tanso *58*, 246 (1969)
180. Fester, G. A., Martinuzzi, E. A., Ricciardi, A. I. A.: Erdöl u. Kohle *10*, 840 (1957)
181. Kroepelin, H., Kadelbach, H., Kopsch, H., Winter, E.: Dechema Monograph *29*, 204 (1957)
182. Tiberio, U., Marcianti, M., Salaidi, G., Piazzi, M.: Ricerca Sci. *27*, 3063 (1957); *29*, 116 (1959)
183. Luther, H., Mertlich, K.: Naturwissenschaften *45*, 57 (1958)
184. Smol'yaninov, S. I., Kravstov, A. V., Vishnevetskii, I. I., Kotlova, L. F., Danilova, L. F.: Sb. Tr. Moldykh. Uch., Tomsk. Politekh. Inst. (1), 138 (1973)
185. Smol'yaninov, S. I., Kravstov, A. V., Vishnevetskii, I. I., Kotlova, L. F., Kiryaeva, A.: Mater. Obl. Nauchn. Konf. Vses. Khim. O-va., Posvyashch. 75- Letiju Khim. Teknol. Fak. Tomsk. Politekh Inst., 3rd, 10 (1972)
186. Novikov, I. K.: Zh. Prikl. Khim. *38*, 1332 (1965)
187. Novikov, I. K., Kozlov, Yu. A.: Neftepererab. Neftekhim. (2), 149 (1967)
188. Sokol'skii, D. V., Buvalkina, L. A., Morozov, L. G., Shulyar, B. N.: Izv. Akad. Nauk Kaz. SSR Ser. Khim. *18* (4), 61 (1968)
189. Morozov, L. G., Buvalkina, L. A., Solkol'skii, D. V.: Khim. Khim. Tekhnol. (11), 127 (1971)
190. Morozov, L. G., Buvalkina, L. A., Sokol'skii, D. V.: Dokl. Vses. Konf. Khim. Atsetilena, 4th, (3), 397 (1972)
191. Semenov, L. V., Davydov, V. P., Chukanova, O. M.: Neftepererabotka i Neftekhim., Nauchn-Tekhn. Sb. (2), 29 (1964)
192. Shishakov, N. V., Topol'skaya, F. M.: Neftepererabotka i Neftekhim., Nauchn-Tekhn. Sb. (10), 28 (1964)
193. Shishakov, N. V., Topol'skaya, F. M., Khotuntsev, L. L., Rapiovets, L. S., Teoriya i Teknol. Protsessov Pererabotki Topliv, Inst. Goryuch. Iskop. 93 (1966)
194. Pasman, H. J., Vlugter, J. C., Brenkinte, C. J.: Brennstoff-Chem. *46*, 358 (1965)
195. Nagata, K.: Kogyo Kagaku Zasshi *68*, 1807 (1965)
196. Iosebidze, D. S., Melikadze, L. D.: Soobshch. Akad. Nauk Gruz. SSR *47* (1), 61 (1967)
197. Karn, F. S., Friedel, R. A., Sharkey, Jr., A. G.: Chem. Ind. (Lond.), 239 (1970)

198. Coffman, J. A., Browne, W. R.: Scient. Am. *212* (6), 91 (1965)
199. Allt, P. K., Datta, P., Macey, W. A. T., Semmens, B.: J. Appl. Chem. *18*, 213 (1968)
200. Suhr, H.: Acta. Cient. Venez. *27* (4), 159 (1976)
201. Martin, R. L., Grant, J. A.: Anal. Chem. *37*, 644 (1965)
202. Thompson, C. J., Coleman, H. J., Hopkins, R. L., Rall, H. T.: Spec. Tech. Publs. Am. Soc. Test. Mater. (389), 329 (1965)
203. Berthelot, M.: Compt. Rend. *82*, 1357 (1876)
204. Losanitsch, S. M.: Ber. *42*, 4394 (1909)
205. Klabunde, K. J., Skell, P. S.: J. Amer. Chem. Soc. *93*, 3807 (1971)
206. Moscow Inst. of Chem. Eng.: Japanese Pat. 7,687,493 (July 31, 1976)
207. Fjeldstad, P. E., Undheim, K.: Acta Chem. Scand. *B30*, 375 (1976)
208. Suhr, H., Henne, P.: Liebigs Ann. Chem., 1610 (1977)
209. Henne, P.: Paper presented at the Third Internat. Symp. Plasma Chem., Limoges (July 1977)
210. Yelm, K. E., Venugopalan, M.: Unpublished
211. Isomura, Y.: J. Electrochem. Soc. Japan *17*, 69 (1949)
212. Unschuld, H. M., Alther, J. G.: U.S. Pat. 2,504,058 (1950)
213. Mitsumi, S., Takihara, T.: Japanese Pat. 7,233,073 (November 16, 1972)
214. Inoue, K.: Japanese Pat. 7,445,281 (December 3, 1974); 7,445,682 (December 5, 1974)
215. Kawawna, Y., Makino, M.: Japanese Pat. 7,354,105 (1973)
216. Reggel, L., Blaustein, B. D., DelleDonne, C. L., Friedman, S., Steffgen, F. W., Winslow, J. C.: Fuel *55* (3), 170 (1976)
217. Encyclopedia of chemical technology, 2nd ed., Vol. 5, Kirk, R. E., Othmer, D. F. (ed.). p. 627. New York: Wiley 1964
218. Hill, G. R.: Chem. Tech., 294 (1972)
219. Whitehurst, D. D.: ACS Symp. Ser. *71*, 1 (1978)
220. Wiser, W.: Preprints Fuel Div. ACS Meeting *20* (2), 122 (1975)
221. Sanada, Y., Berkowitz, N.: Fuel *48*, 375 (1969)
222. Kobayashi, K., Berkowitz, N.: Fuel *50*, 254 (1971)
223. Scott, T. W., Venugopalan, M.: Nature *262*, 48 (1976)
224. Scott, T. W., Chu, K.-C., Venugopalan, M.: Chem. Ind. (Lond.), 739 (1976)
225. Venugopalan, M.: Unpublished
226. Dzhamanbaer, A. S., Ibravev, S. O.: Freiberg Forschungs. A, A577, 93 (1977)
227. Fu, Y. C., Blaustein, B. D., Wender, I.: Chem. Eng. Progr. Symp. Ser. *67*, 47 (1971)
228. Nishida, S., Berkowitz, N.: Fuel *52*, 262 (1973)
229. Fu, Y. C., Blaustein, B. D.: Chem. Ind. (Lond.), 1257 (1967)
230. Fu, Y. C., Blaustein, B. D.: Fuel *47*, 463 (1968)
231. Fu, Y. C., Blaustein, B. D.: Ind. Eng. Chem., Process Design Develop. *8* (2), 257 (1969)
232. Fu, Y. C., Blaustein, B. D., Sharkey, Jr., A. G.: Fuel *51*, 308 (1972)
233. Nishida, S., Berkowitz, N.: Fuel *52*, 267 (1973)
234. Ammann, P. R., Baddour, R. F., Johnston, M. M., Mix, T. W., Timmins, R. S.: Chem. Eng. Progr. *60*, 52 (1964)
235. Kawana, Y., Makino, M., Kimura, T.: Kogyo Kagaku Zasshi *69*, 1144 (1966)
236. Krukonis, V. J., Gannon, R. E., Modell, M.: Adv. Chem. Ser. *131*, 29 (1974)
237. Krukonis, V. J., Gannon, R. E., Schoenberg, T.: Can. Eng. Conf., 19th, Edmonton (1969)
238. Razina, G. N., Fedoseev, S. D., Gvozdarev, V. G., Staroverov, V. A.: Tr. Mosk. Khim.-Tekhnol. Inst. *91*, 94 (1976)
239. Ladner, W. R., Wheatley, R.: Fuel *50*, 443 (1971)
240. Bond, R. L., Galbraith, I. F., Ladner, W. R., McConnell, G.I.T.: Nature *200*, 1313 (1963)
241. Bond, R. L., Ladner, W. R., McConnell, G.I.T.: Fuel *45*, 381 (1966)
242. Bond, R. L., Ladner, W. R., McConnell, G.I.T.: Adv. Chem. Ser. *55*, 650 (1966)
243. Graves, R. D., Kawa, W., Hiteshue, R. W.: Ind. Eng. Chem., Process Design Develop. *5*, 59 (1966)
244. Kawana, Y., Makino, M., Kimura, T.: Kogyo Kagaku Zasshi *69*, 1144 (1966); Intern. Chem. Eng. *7*, 359 (1967)
245. Kawana, Y., Makino, M.: Kogyo Kagaku Zasshi *70*, 1657 (1967)

246. Klan, J., Dolal, V.: Acta Mont. *39*, 49 (1976)
247. Kulczycka, J.: Khim. Tverd. Topl. (Moscow), (3), 102 (1978)
248. Hebecker, D., Moegel, G., Heilemann, U.: East German Pat. 114,395 (August 5, 1975)
249. Srivastava, S. K., Chakravartty, S. C., Dixit, L. P.: Chem. Era *13*, 368 (1978)
250. Kulczycka, J.: Pr. Gl. Inst. Gorn., Komun., No. 549 (1972)
251. Chakravartty, S. C., Dutta, D., Lahiri, A.: Fuel *55*, 43 (1976)
252. Littlewood, K.: Symp. on Chemicals and Oil from Coal, Dhanbad, India, 517 (1969)
253. Nicholson, R., Littlewood, K.: Nature *236*, 397 (1972)
254. Chakravartty, S. C., Dutta, D.: Fuel *55*, 254 (1976)
255. Archbold, E., Hughes, T. P.: Nature *204*, 670 (1964)
256. Archbold, E., Harper, D. W., Hughes, T. P.: Brit. J. Appl. Phys. *15*, 1321 (1964)
257. Nelson, L. S.: J. Phys. Chem. *63*, 433 (1959); Science *136*, 296 (1962)
258. Hawk, C. O., Schlesinger, M. D., Hiteshue, R. W.: Rep. Invest. U.S. Bur. Min. no. 6264 (1963)
259. Rau, E., Seglin, L.: Fuel *43*, 147 (1964)
260. Sharkey, Jr., A. G., Shultz, J. L., Friedel, R. A.: Nature *202*, 988 (1964); Adv. Chem. Ser. *55*, 643 (1966)
261. Granger, A. F., Ladner, W. R.: Combustion and Flame *11*, 518 (1967); Fuel *49*, 17 (1970)
262. McIntosh, M. J.: Fuel *55*, 59 (1976)
263. Green, N. W., Duraiswamy, K., Lumpkin, R. E., Knell, E. W., Mirza, Z. I., Winter, B. L.: U.S. Pat. 4,085,030 (April 18, 1978)
264. Hayward, R. J. R., Ladner, W. R., Wheatley, R.: Fuel *49*, 223 (1970)
265. Rau, E., Eddinger, R. T.: Fuel *43*, 246 (1964)
266. Bond, R. L., Ladner, W. R., Wheatley, R.: Fuel *47*, 213 (1968)
267. Petrakis, L., Grandy, D. W., Anal. Chem. *50*, 303 (1978)
268. Karn, F. S., Friedel, R. A., Sharkey, Jr., A. G.: Carbon *5*, 25 (1967)
269. Shultz, J. L., Sharkey, Jr., A. G.: Carbon *5*, 57 (1967)
270. Karn, F. S., Friedel, R. A., Sharkey, Jr., A. G.: Fuel *48*, 297 (1969)
271. Karn, F. S., Friedel, R. A., Sharkey, Jr., A. G.: Fuel *51*, 113 (1972)
272. Karn, F. S., Singer, J. M.: Fuel *47*, 235 (1968)
273. Karn, F. S., Sharkey, Jr., A. G.: Fuel *47*, 193 (1968)
274. Joy, W. K., Ladner, W. R., Pritchard, E.: Nature *217*, 640 (1968)
275. Joy, W. K., Ladner, W. R., Pritchard, E.: Fuel *49*, 26 (1970)
276. Kawana, Y.: Shinku Kagaku *15* (2), 56 (1967)
277. Sanada, Y.: Nenryo Kyokai-Stu. *49*, 580 (1970)
278. Klan, J., Nemec, J.: Acta Mont. *39*, 7 (1976)
279. Ponnik, Yu. A.: Khim. Tverd. Topl. *3*, 151 (1972)
280. Gebert, F.: German Pat. 2,062,091 (June 1972)
281. Bal, S., Swierezek, R., Musialski, A.: Koks, Smola, Gaz. *16* (4), 96 (1971)
282. Kawana, Y.: Chem. Econ. Eng. Rev. *4* (1), 13 (1972)
283. Tylko, J. K., Skinner, G. F.: Paper presented at the Third Internat. Symp. Plasma Chem., Limoges (July 1977)
284. Noda, H., Hiratake, S.: Japanese Pat. 7,814,702 (February 9, 1978)
285. Szuba, J., Wasilewski, P., Mikolasjska, M., Bal, S., Surenczek, R.: Polish Pat. 83,271 (December 15, 1976)
286. Maugh II, T. H.: Science *178*, 45 (1972). See also Adv. Chem. Ser. *131* (1974); *146* (1975)
287. Perry, H. S.: Scient. Am. *230*, 19 (1974)
288. Blaustein, B. D., Fu, Y. C.: Adv. Chem. Ser. *80*, 259 (1969)
289. Fischer, F., Peter, K.: Brennstoff-Chem. *12*, 268 (1931)
290. Lunt, R. W.: Proc. Roy. Soc. (Lond.) *108A*, 172 (1925)
291. McTaggart, F. K.: Australian Pat. Appln. No. 55,010/59 (1959)
292. LeRoy, D. J., Steacie, E. W. R.: J. Chem. Phys. *12*, 369 (1944)
293. Deurbrouck, A. V.: Rep. Invest. U.S. Bur. Mines, 7633 (1972)
294. Gluskoter, H. J.: Preprints Fuel Div. ACS Meeting *20* (2), 94 (1975)
295. Boateng, D. A. D., Phillips, C. R.: Fuel *55*, 318 (1976)
296. Attar, A., Corcoran, W. H.: Ind. Eng. Chem., Prod. Res. Dev. *16* (2), 168 (1977)

297. Attar, A.: Fuel *57*, 201 (1978)
298. Coal desulfurization. ACS Symp. Ser. *64* (1977)
299. Attar, A., Corcoran, W. H., Gibson, G.: Preprints Fuel Div. ACS Meeting *21* (7), 106 (1976)
300. Maa, P. S., Lewis, C. R., Hamrin, C. E.: Fuel *54*, 62 (1975)
301. Robinson, L.: Fuel *55*, 193 (1976)
302. For review of early work see Glockler, G., Lind, S. C.: The electrochemistry of gases and other dielectrics. pp. 222–223. New York: Wiley 1939
303. Maruzen Oil Co., Ltd.: Japanese Pat. 14,413 (July 9, 1965)
304. Stokes, C. S.: Adv. Chem. Ser. *80*, 390 (1969); in: Reactions under plasma conditions, Vol. II. Venugopalan, M. (ed.). p. 290. New York: Wiley-Interscience 1971
305. Zavitsanos, P. D., Bleiler, K. W.: German Pat. 2,657,472 (June 1977)
306. Block, S. S., Sharpe, J. B., Darlage, L. J.: Fuel *54*, 113 (1975)
307. Sinha, R. K., Walker, Jr., P. L.: Fuel *51*, 125, 329 (1972)
308. Steinberg, M., Yang, R. T., Hom, T. K., Berlad, A. L.: Fuel *56*, 227 (1977)
309. Amberg, C. H.: Can. J. Chem. *41*, 1966 (1963)
310. Frye, C. G., Mosby, J. F.: Chem. Eng. Prog. *63*, 66 (1967)
311. Givens, E. N., Venuto, P. B.: Preprints Petro. Chem. Div. ACS Meeting *15* (4), A183 (1970)
312. Hartough, H. D.: Thiophene and its compounds. New York: Interscience 1952
313. Fu, Y. C.: Chem. Ind. (Lond.), 876 (1971)

Received April 12, 1979

Kinetics of Dissociation Processes in Plasmas in the Low and Intermediate Presssure Range

Mario Capitelli and Ettore Molinari

Istituto di Chimica Generale ed Inorganica dell'Università and Centro di Studio per la Chimica dei Plasmi del C.N.R., Via Amendola 173, I-70126 Bari, Italy
Istituto di Chimica Generale ed Inorganica dell'Università, I-00100 Roma, Italy

Table of Contents

1 Introduction

The problems involved in modelling weakly ionized plasmas in molecular gases, operated at pressures below about 100 mbar and with average electron energies in the range 0.5–3 eV, are typically those of non-equilibrium systems.

Great progress has been achieved in the description of this type of plasmas since the operation of the first gas laser[1].

Electron energy distribution functions are generally non-Maxwellian, their form reflecting the dominant features of the electron molecule energy exchange processes. Electron molecule collision rates are averages of cross sections over the distribution of electron energies and evaluation of these quantities requires therefore a knowledge of the actual electron distribution function (edf). When energy is pumped by the electric field into the vibrational system of the molecule by e-v, r excitation processes, this energy is distributed via v-v (vibration-vibration), v-r (vibration-rotation), v-t (vibration-translation), r-r (rotation-rotation) and r-t (rotation-translation) collisional energy transfer processes.

The result of combined excitation by electrons and collisional energy transfer is the creation of a nonequlibrium situation in which the electrons and the different molecular degrees of freedom are characterized by different temperatures. In general, $T_e > T_v > T_r > T_g$. The use of temperature to characterize distributions of electronic, vibrational, rotation and kinetic energies is however meaningful for Boltzmann distributions only. While, in general, rotational and kinetic energies follow Boltzmann distributions, large deviations from this distribution have been observed for both electronic and vibrational energies. High vibrational levels can be overpopulated and the excess energy is fed back to the electrons via superelastic collisions, with a consequent modifications of the edf. This complex non-equilibrium situation controls the operation of gas lasers and this justifies the large attention devoted to these problems.

In the present chapter the attention will be focussed on processes of molecular dissociation under nonequilibrium plasma conditions.

It will be shown that, depending on the system and on the relevant discharge parameters, a variable but generally large fraction of the energy pumped into the type of plasma under consideration is transferred to the vibrotational system of the molecule which can then dissociate via a ladder-climbing mechanism across the vibrotational manifold of the ground electronic state. This represents an additional mechanism which should be examined in conjunction with dissociation processes involving the direct electron excitation to an electronically excited state of the molecule which then dissociates or predissociates

$$M_2 + e \longrightarrow M_2^* + e \longrightarrow 2M + e$$

The distribution of the population of vibrational levels are either non-Boltzmann or, in any case, characterized by vibrational temperatures $T_v > T_g$. One important consequence is that the excited state M_2^* can now be pumped by electrons not only from the $v = 0$ level of the ground state, as was generally assumed, but also from $v > 0$ levels, with an obvious increase of the pumping rate from ground state to M_2^*.

The relative importance of the various contribution to the plasmochemical dissociation rate strongly depends on the system and on plasma conditions.

When the molecule dissociates atoms are produced and some interesting feedback processes should be considered.

The parent atom is, in general, very effective in removing vibrational energy by v-t transfer processes. As the concentration of atoms builds up the distribution of vibrational levels is progressively changed into a Boltzmann distribution with T_v approaching T_g. As a consequence, dissociation processes controlled by the vibrational distribution will be progressively depressed. However, the increased atom concentration increases the homogeneous three body recombination, a process which tends to overpopulate upper vibrational levels. The vibrational distribution is thus modified with a plateau appearing at high v's, a situation which favours dissociation via the vibrational ladder.

An important consequence of the presence of the atoms is that they modify the electron distribution function in the sense that, for a given reduced field E/N ($V cm^2$), the average electron energy becomes higher. This will favour processes with a high energy threshold, such as dissociation or ionization by direct electron impact.

In the sections to follow an attempt will be made to describe a model containing the essential features outlined above. A comparison with experimental findings will be made, when possible.

A well suited *case study* is represented by the dissociation of molecular hydrogen, which will form the subject matter of Sect. 2.

Section 3 illustrates the distinct features which characterize the dissociation of different diatomic molecules (N_2, O_2, CO, HF).

Section 4 will be devoted to polyatomic molecules. A selection of papers is made to illustrate the point of view of the authors on the influence of non equilibrium distributions on the rates of plasmochemical dissociation reactions. The material presented has mostly been derived from the authors' group investigations directly related to it.

2 A Case Study:
The Dissociation of Molecular Hydrogen

2.1 Evaluation of the Electron Energy Distribution Function (edf)

Electron energy distribution functions have been calculated by different authors[2, 3, 34–36, 68–70] by solving the appropriate Boltzmann equation.

The results reported in Fig. 1–6, have been obtained by using the formulation of Ref.[69].

Figure 1 shows the edf calculated for H_2 by solving the relevant Boltzmann equation with the inclusion of the following inelastic processes, besides the elastic ones[2].

a) Vibrational excitation

$$e + H_2 \ (v = 0) \longrightarrow e + H_2 \ (v = 1, 3) \tag{I}$$

M. Capitelli and E. Molinari

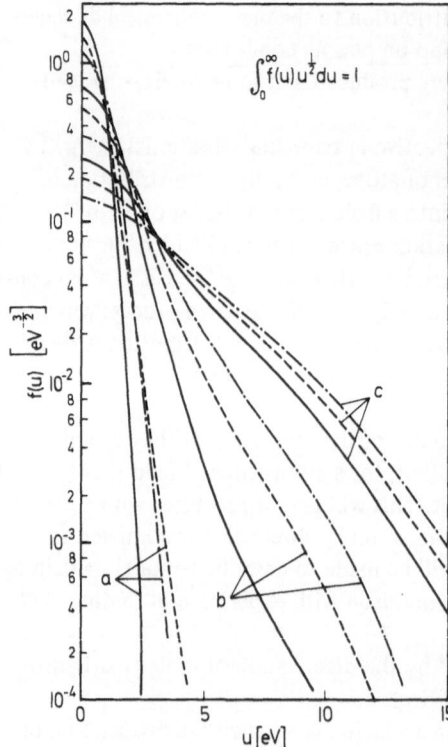

Fig. 1. Electron energy distribution functions $f(u)$ as a function of energy at different E/N values (——: "cold" molecular gas; – – –: vibrationally excited molecular gas; –·–·–: H_2/H 1:1 mixture) (*a* E/N = 10^{-16} V cm^2; *b* E/N = 3 × 10^{-16} V cm^2; *c* E/N = 6 × 10^{-16} V cm^2) (From Ref.[2])

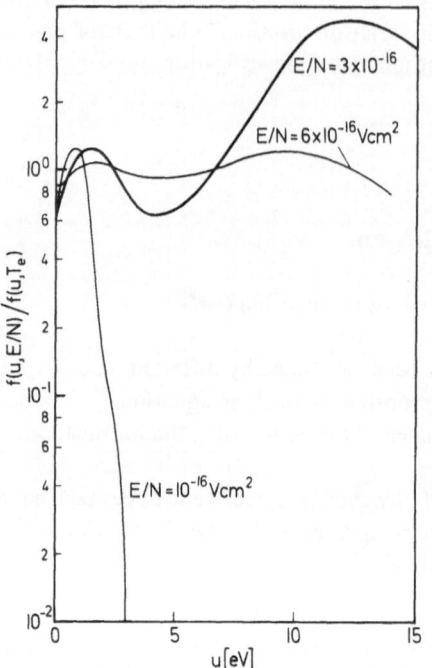

Fig. 2. Values of $f(u, E/N)/f(u, T_e)$ as a function of energy at different E/N values for a cold H_2 molecular gas. $f(u, T_e)$ represents the Maxwell approximation at $\bar{u}_r \equiv KT_e$ (From Ref.[2])

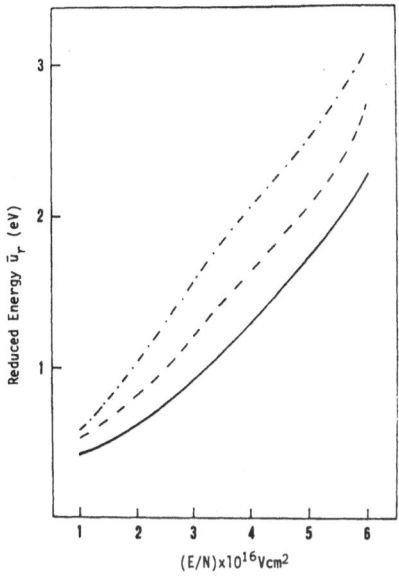

Fig. 3. Reduced averaged electron energies $\bar{u}_r$ as a function of E/N (—— "cold" molecular gas; – – – vibrationally excited molecular gas; –·–·– H_2/H 1:1 mixture) (From Ref.[2])

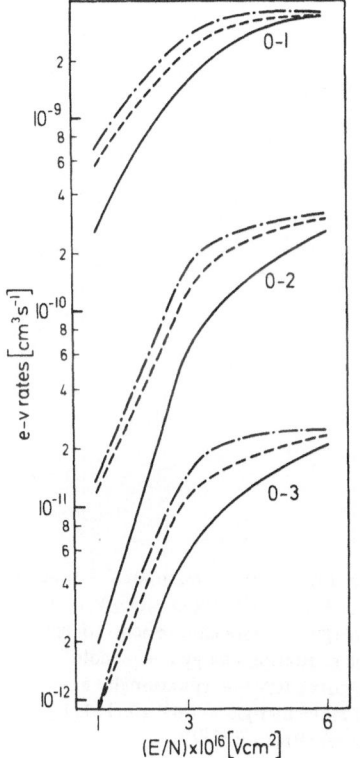

Fig. 4. *e-v* (electron vibration) rates in the ground electronic state as a function of E/N (—— "cold" molecular gas; – – – vibrationally excited molecular gas; –·–·– H_2/H 1:1 mixture) (From Ref.[2])

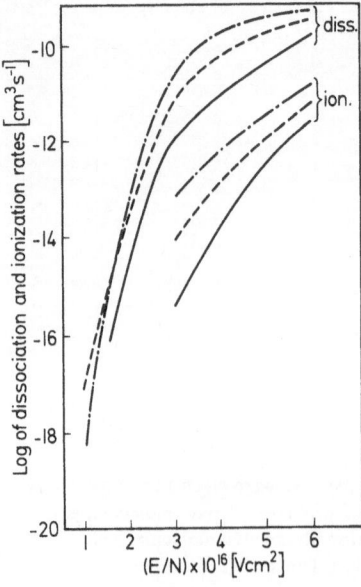

Fig. 5. Dissociation and ionization rates as a function of E/N (—— "cold" molecular gas; – – – vibrationally excited molecular gas; –·–·– H_2H 1:1 mixture) (From Ref.[2])

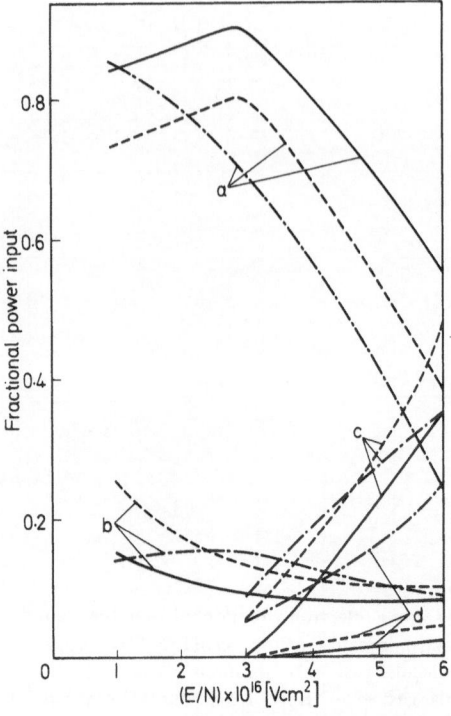

Fig. 6. Fractional power input as a function of E/N (*a* vibrational energy; *b* elastic energy; *c* dissociation energy; *d* electronic excitation energy (—— "cold" molecular gas; – – – vibrationally excited molecular gas; –·–·– H_2/H 1:1 mixture) (From Ref.[2])

b) Dissociation

$$e + H_2 \, (v = 0) \longrightarrow e + H_2^* \longrightarrow e + 2H \tag{II}$$

c) Electronic excitation

$$e + H_2 \, (v = 0) \longrightarrow e + H_2^* \tag{III}$$

d) Ionization

$$e + H_2 \, (v = 0) \longrightarrow e + H_2^+ + e \tag{IV}$$

One notices that 1) all processes start from the ground vibro-electronic state of H_2. 2) rotational excitation and thermalization effects of electron- electron interactions have not been included. In fact rotational excitation is important at values of the reduced field E/N lower than those considered in the calculations and electron-electron interactions can be neglected in weakly ionized plasmas (ionization degree $z_i \lesssim 10^{-5}$).

Figure 1 (full lines) shows that, at $E/N \simeq 10^{-16}$ $V cm^2$, the edf is characterized by the absence of electrons with energies greater than $3 eV$, a consequence of the rapid rise of the vibrational cross-section in the $1-3$ eV energy range. This effect becomes less marked at higher E/N, where the electrons can penetrate the vibrational barrier.

The non-maxwellian character of the calculated distribution can be appreciated from Fig. 2, where the ratio $f(u, E/N)/f(u, T_e)$ has been plotted as a function of electron energy u, with T_e (eV) taken equal to the reduced average energy

$$\overline{u}_r = \frac{2}{3} \, \overline{u}.$$ Rates of vibrational excitation (e-v), of dissociation and of ionization can

be easily obtained from the calculated edf, together with the reduced average energy. Figures 3–5 illustrate the dependence of these quantities on E/N. The high threshold energies of dissociation and ionization, which effectively overlap the high energy tail of the edf, are responsible for the rapid increase with E/N of the rates of the two processes.

Edf's of Fig. 1 (full lines) do not include the effects of vibrationally excited molecules and of H-atoms.

In the presence of an appreciable population of vibrationally excited molecules an additional process should be included in the Boltzmann equation, namely

e) Superelastic vibrational collisions

$$e + H_2 \, (v = 1, 3) \longrightarrow e + H_2 \, (v = 0) \tag{V}$$

Only levels connected to $v = 0$ have been considered for which cross sections are available. Edf's including process V are given as dashed lines in Fig. 1.

The tails of these electron energy distributions benefit of the presence of excited vibrational states which, in this calculation, have been assumed to obey a Boltzmann distribution with $T_v = 3500$ K. The tails of these distributions are more populated, by orders of magnitude than the corresponding ones for a vibrationally "cold" gas. As a consequence, processes controlled by the edf tail, such as dissociation and ionization, will strongly be affected by vibrational disequilibrium, while a minor influence is expected on e-v rates and $\bar{u}_r$ which depend more on the bulk of the edf. (Fig. 1, 3–5).

In the presence of hydrogen atoms the Boltzmann equation should be solved for a mixture of the two species (H_2, H)[3]. Elastic and inelastic processes involving H-atoms should now be considered. The following have been included in the calculations of Ref.[2].

$$e + H(1s) \longrightarrow e + H(2s, 2p) \tag{VI}$$

$$e + H(1s) \longrightarrow e + H^+ + e \tag{VII}$$

The results obtained for an equimolar mixture of vibrationally "cold" H_2 and H-atoms, have been reported in figure 1 as dashed dotted lines. Edf tails again benefit, owing to the absence of vibrational losses in the atomic system. The presence of atoms can therefore increase the rates of dissociation and ionization by one or two orders of magnitude (Fig. 5). A minor influence is again expected on e-v rates, and $\bar{u}_r$ (Figs. 3–4).

In the course of a dissociation process in a plasma, the rate constant of this process increases with time as a consequence of the increased concentration of atoms and of the increased vibrational temperature, both affecting the edf's tail.

In the calculations presented above T_v and the concentration of H-atoms have been included as independent quantities in order to illustrate their effects on the edf. In a rigorous treatment the Boltzmann equation for the electron distribution function should be coupled to the equations which describe the temperal evolution of vibrational level densities and atom concentrations (see below).

2.2 Fractional Power Transfer

The balance of the electron energy can be written in implicit form as

$$\dot{E}_f = \dot{E}_{elast} + \dot{E}_{vibr} + \dot{E}_{electr} + \dot{E}_{dis} + \dot{E}_{ion} \tag{1}$$

were $\dot{E}_f$ is the rate of energy gained by the electrons from the DC field and the right hand side of Eq. (1) gives the rate of energy losses due to elastic and inelastic processes (vibrational, electronic, dissociation, ionization processes). It is of particular interest for both laser operation and plasmochemical processes to evaluate the relative importance of the various terms. Typical results have been reported in Fig. 6 as a function of E/N for a vibrationally excited gas with $T_v = 3500$ K, and a 1/1 mixture of vibrationally cold H_2 and H-atoms.

It should be appreciated that at $E/N < 5 \times 10^{-16}$ V cm^2 energy is almost entirely deposited in the vibrational system of H_2. The marked influence of vibrational excitation and of the atoms on fractional power deposition emerges quite clearly from this figure.

2.3 Dissociation Mechanisms of Molecular Hydrogen

2.3.1 The Pure Vibrational Mechanism (PVM)

According to this mechanism[4-8], electron molecule collisions populate the low vibrational levels of the ground electronic state of H_2 via e-v processes

$$e + H_2(v) \rightleftarrows e + H_2(i) \tag{VIII}$$

The vibrational quanta are then redistributed over higher vibrational levels by means of v-v (vibration-vibration) energy exchanges

near resonant $\quad H_2(v) + H_2(v) \rightleftarrows H_2(v+1) + H_2(v-1)$ (IX)

non resonant $\quad H_2(1) + H_2(v) \rightleftarrows H_2(0) + H_2(v+1)$ (X)

On the other hand v-t (vibration-translation) energy transfer tends to depopulate the $v+1$ level

$$H_2(v+1) + H_2 \rightleftarrows H_2(v) + H_2 \tag{XI}$$

Dissociation occurs when vibrational quanta cross a pseudolevel or dissociation level $(v'+1)$ located at the top of the vibrational ladder $(v' = 14$ is the last bound vibrational level of $H_2({}^1\Sigma))$.

The dissociation process can be written as

$$H_2(v') + H_2 \xrightarrow{P_{v',v'+1}} H_2(v'+1) + H_2 \longrightarrow 2H + H_2 \tag{XII}$$

$$H_2(v') + H_2(v) \xrightarrow{P_{v,v-1}^{v',v'+1}} H_2(v'+1) + H_2(v-1) \longrightarrow 2H + H_2(v-1) \tag{XIII}$$

The dissociation rate r_d is then written as the rate of population of the pseudolevel in the form

$$r_d = dN_{v'+1}/dt = N_{H_2}N_{v'}P_{v',v'+1} + N_{v'}\sum_0^{v'} N_v P_{v,v-1}^{v',v'+1} = k_d(t)\sum_0^{v'} N_v \tag{2}$$

where $P_{v',v'+1}$ and $P_{v,v-1}^{v',v'+1}$ represent the rate coefficients of processes (XII) and (XIII), respectively $(cm^{-3}s^{-1})$. Equation (2) is then coupled to a system of v' vibrational master equations, each of which describes the temporal evolution of the population density of the vth vibrational level submitted to the action of e-v, v-v and v-t energy transfers[4-8]

$$dN_v/dt = (dN_v/dt)_{e-v} + (dN_v/dt)_{v-v} + (dN_v/dt)_{v-t} \tag{3a}$$

$$(dN_v/dt)_{e-v} = n_e \sum_{i=0}^{v'} (k_{iv}^e N_i - k_{vi}^e N_v) \tag{3b}$$

$$(dN_v/dt)_{vv} = \sum_{i=1}^{v'+1} P^{i-1,i}_{v+1,v} \{N_{v+1}N_{i-1} - N_vN_i \exp - [(E_{v+1} + E_{i-1} - E_v - E_i)/kT_g]\} +$$

$$- \sum_{i=0}^{v'} P^{i,i+1}_{v,v-1} \{N_vN_i - N_{v-1}N_{i+1} \exp - [(E_v + E_i - E_{i+1} - E_{v-1})/kT_g]\}$$

$$(3c)$$

$$(dN_v/dt)_{v\text{-}t} = N_{H_2}P_{v+1,v} \{N_{v+1} - N_v \exp - [(E_{v+1} - E_v)/kT_g)]\} +$$

$$- N_{H_2}P_{v,v-1} \{N_v - N_{v+1} \exp - [(E_v - E_{v-1})/kT_g]\}$$

$$(3d)$$

After an induction time, the phenomenological pseudo-first order dissociation constant $k_d(t)$ (Eq. 2) settles to a stationary value k^s_d (s^{-1}), which can be taken as a measure of the contribution of PVM to dissociation (see Ref.[4]). Values of k^s_d and of the corresponding N_v distributions have been reported as a function of gas temperature T_g, pressure p and electron density n_e, at different E/N in Figs. 7–11[9]. The sharp decrease of k^s_d with increasing T_g (Fig. 7), a result opposite to that normally observed under thermal conditions ($n_e = 0$), is caused by the higher rates of the deactivating v-t processes which tend to thermalize the N_v distributions. Moreover, at high T_g, the net rates of up-pumping by the v-v mechanism decrease as a consequence of the increasing importance of the reverse processes. N_v distributions, for conditions reported in Fig. 7, are given in Fig. 8. Some comments on the shape of these distributions are appropriate:

If the system of vibrational master equations (vme) is solved without the inclusion of v-v and v-t processes, i.e. one considers e-v processes only, the N_v distribution is well represented by a Boltzmann distribution at the electron temperature T_e for all levels connected by e-v processes.

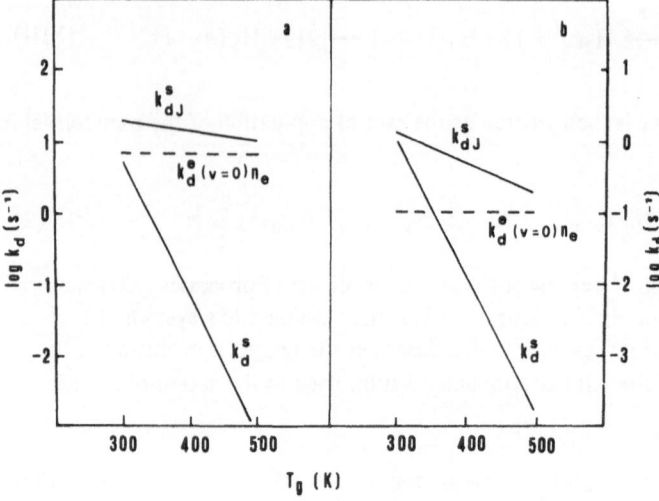

Fig. 7a and b. Values of the dissociation constants (s^{-1}) as a function of gas temperature ($N_{H_2} = 10^{17}$ cm^{-3}, $n_e = 10^{12}$ cm^{-3}) (a E/N = 3 x 10^{-16} V cm^2; b E/N = 2 x 10^{-16} V cm^2) (From Ref.[9])

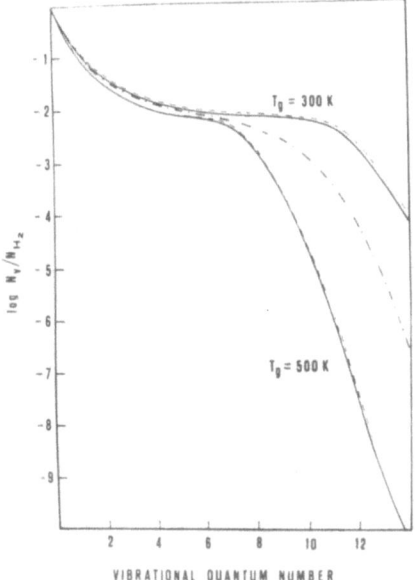

Fig. 8. Vibrational populations densities ($\log N_v/N_{H_2}$) as a function of vibrational quantum number at two different gas temperatures ($N_{H_2} = 10^{17}$ cm^{-3}, $n_e = 10^{12}$ cm^{-3}, E/N = 3 x 10^{-16} cm^2; t = 10^{-3} s) (—— JVE; ––– PVM; –·–·– JVE with recombination). (From Ref.[9])

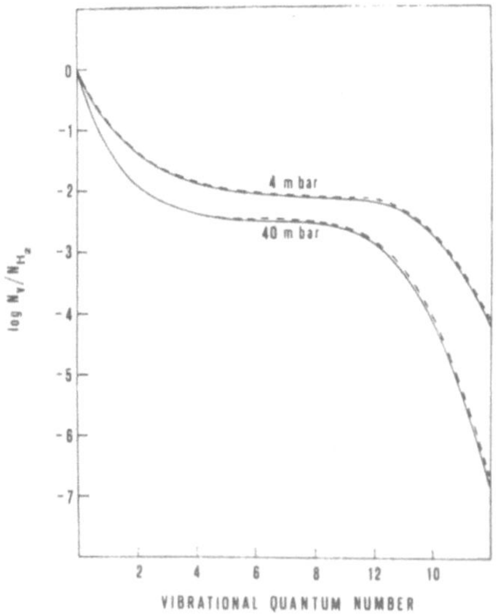

Fig. 10. Vibrational population densities ($\log N_v/N_{H_2}$) as a function of vibrational quantum number at two pressures ($n_e = 10^{12}$ cm^{-3}, $T_g = 300$ K, E/N = 3 x 10^{-16} Vcm^2, t = 10^{-3} s) (—— JVE; ––– PVM) (From Ref.[9])

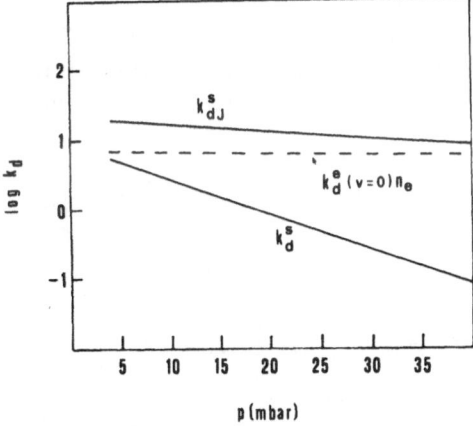

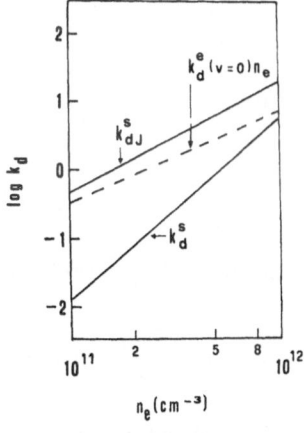

Fig. 9. Values of the dissociation constants (s^{-1}) as a function of pressure ($n_e = 10^{12}$ cm^{-3}, $T_g = 300$ K, E/N = 3 x 10^{-16} Vcm^2) (From Ref.[9])

Fig. 11. Values of the dissociation constants (s^{-1}) as a function of electron density (E/N = 3 x 10^{-16} Vcm^2, $T_g = 300$ K, p = 4 mbar) (From Ref.[9])

When the system of vme is solved by retaining v-v terms only, under an initial disequilibrium between the ground and the first vibrational level, defined by a vibrational temperature $\theta_1 = E_{10}/k\ln(N_0/N_1)$, one obtains a Treanor distribution in the form[10]

$$N_v/N_{v+1} = \exp(E_{10}/\theta_1 - 2E_{10}\,\delta\,v/T_g) \qquad (4)$$

This type of distribution, which presents an inversion at a vibrational quantum number given by $v_I = T_g/2\delta\,\theta_1 + 0.5$ predicts a high population of excited vibrational levels for $\theta_1/T_g \gg 1$. [In Eq. (4) E_{10} is the energy of level 1 and δ is the anharmonicity of the molecule].

Inclusion in the vme of v-t terms only produces a Boltzmann distribution at the gas temperature T_g. The inclusion of e-v, v-v and v-t terms produces the distributions of Fig. 8: e-v processes tend to establish the non equilibrium vibrational temperature $\theta_1 > T_g$. v-v processes, which have rates several orders of magnitude larger than v-t ones, at low vibrational quantum numbers, tend to create a Treanor distribution up to approximately $v = v_1$. Above a given level, v-t processes dominate the v-v ones and determine a Boltzmann tail characterized by a temperature approaching T_g. The plateau, which extends from approximately $v = v_I$ up to the onset of the Boltzmann tail, is connected to the near resonant v-v terms. The dependence of k_d^s on pressure (Fig. 9) can be interpreted along these lines by examining the normalized distributions of Fig. 10. The same arguments apply to the data of Fig. 11.

The possibility for a diatomic molecule to dissociate according to the PVM mechanism outlined above is therefore bound to a "crossing" of favourable conditions for the relevant e-v, v-v and v-t rates. High e-v and v-v rates should be accompanied by small v-t losses. In the case of hydrogen these conditions are fulfilled at low gas temperatures and low pressures.

2.3.2 The Joint Vibroelectronic Mechanism (JVE)

According to this mechanism[9] hydrogen dissociation occurs through processes (XII), XIII as well as through a direct electronic mechanism (DEM) from each vibrational level

$$e + H_2(v) \xrightarrow{k_d^e(v)} e + H_2(b^3\Sigma_u^+) \longrightarrow e + 2H \qquad (XIV)$$

The rate of dissociation can be written as

$$r_d\,(JVE) = \frac{\partial N_{v'+1}}{\partial t} = r_d\,(PVM) + n_e \sum_0^{v'} N_v\,k_d^e(v) = k_{dj}(t)\sum_0^{v'} N_v \qquad (5)$$

where $r_d(PVM)$ [see Eq. (2)] represents the contribution to r_d of the PVM and the second term represents DEM from each vibrational level. DEM from $v = 0$ only has, so far, been considered as the unique electronic mechanism responsible for molecular dissociation in plasmas. This equation should again be coupled to a system of v' mas-

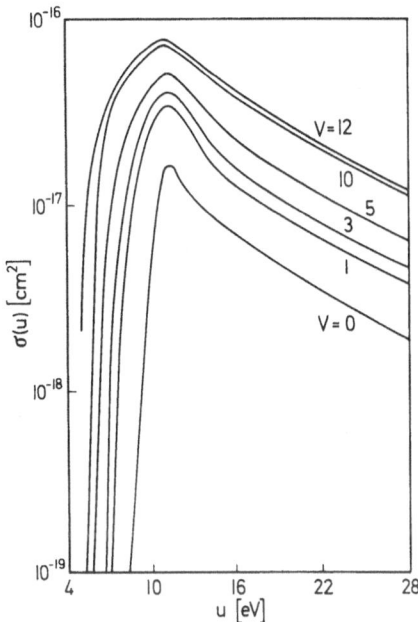

Fig. 12. Cross sections $\sigma(u)$ (cm^2) for the process $e + H_2$ $(^1\Sigma_g, v) \rightarrow e + H_2 (^3\Sigma_u) \rightarrow e + 2H$ as a function of electron energy for different vibrational levels (From Ref.[9])

ter equations with the insertion of the term $n_e N_v k_d^e(v)$ in Eq. (3). In order to calculate $k_d^e(v)$ for processes XIV one needs both the cross sections for each vibrational level and a self consistent edf. Cross sections for processes (XIV), as calculated in Ref.[9] have been reported in Fig. 12 as a function of electron energy for different vibrational levels. With increasing v, threshold energies are lowered, as expected, and the maximum cross section is significantly increased.

The $k_d^e(v)$ rate coefficients have been obtained by using the cross sections of Fig. 12 and the non-Maxwellian electron distribution functions of Fig. 13. The edfs' have been obtained by a numerical solution of the Boltzmann equation (BE) which includes the superelastic vibrational collisions involving the first three vibrational levels, and the dissociation process from all vibrational levels (see Ref.[9] for details). The vibrational population densities inserted in the BE are self-consistent with the quasi-stationary values reported in Figs. 8 and 10. It should be noted that the DEM rates (Fig. 14) depend on E/N as well as on the vibrational non equilibrium present in the discharge, which affects the electron distribution functions, as discussed in Sect. 2.1.

The system of $v' + 1$ differential equations has been solved starting at t = 0 with the initial condition

$$N_v (t = 0) = 0 \quad \text{for } v \neq 0 \tag{6a}$$

$$N_{v=0} (t = 0) = N_{H_2} \tag{6b}$$

After a transient period, the phenomenological pseudofirst order dissociation constant $k_{dj}(t)$ settles to a stationary $k_{dj}^s(s^{-1})$, which can be taken as a measure of the contribution of JVE to dissociation. In the early part of the evolution

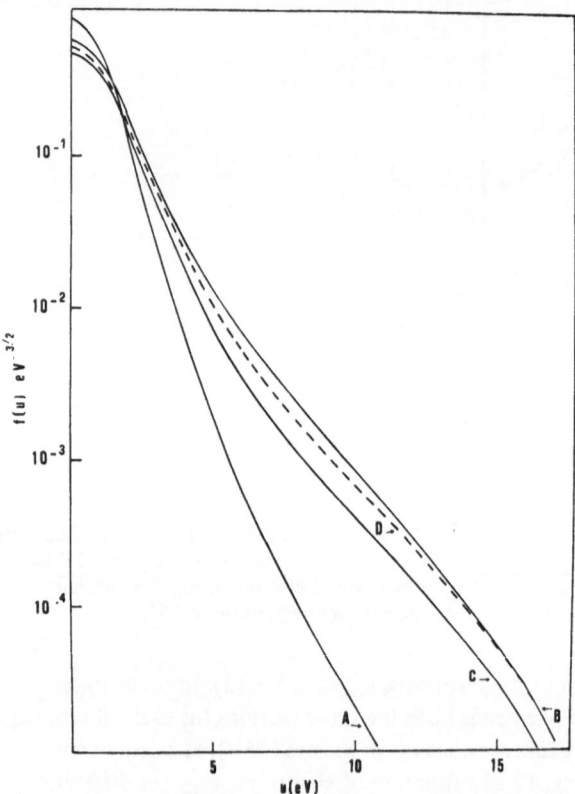

Fig. 13. H_2 electron energy distribution functions $f(u)$ $(eV^{-3/2})$ as a function of energy (eV) selfconsistent with the vibrational distributions at $t = 10^{-3}$ s.

A $E/N = 2 \times 10^{-16}$ V cm^2, p = 4 mbar, T_g = 300 K, $n_e = 10^{12}$ cm^3, u_k = 1.0 eV.

B $E/N = 3 \times 10^{-16}$ V cm^2, u_k = 1.5 eV, $n_e = 10^{12}$; p, T_g as in A.

C $E/N = 3 \times 10^{-16}$ V cm^2, u_k = 1.2 eV, $n_e = 10^{11}$; p, T_g as in A.

D $E/N = 3 \times 10^{-16}$ V cm^2, u_k = 1.3 eV, p = 40 mbar; n_e, T_g as in A (From Ref.[9])

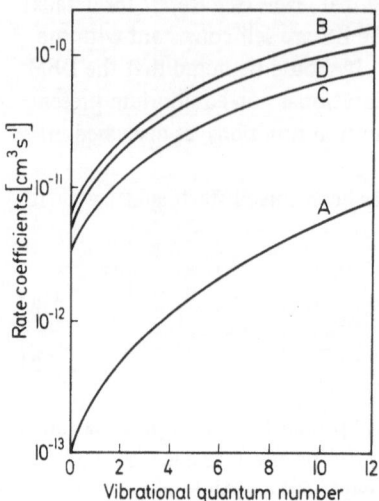

Fig. 14. Rate coefficients (cm^3/s) for process $e + H_2(^1\Sigma_g, v) \longrightarrow e + H_2(^3\Sigma_u) \longrightarrow e + 2H$ as a function of vibrational quantum number for different conditions (A, B, C, D see Fig. 13) From Ref.[9])

$k_{dj}(t) = k_d^e (v = 0) n_e$, while the quasi-stationary value is achieved in times of the order of $(n_e k_{10}^e)^{-1}$.

Figure 7 shows the behaviour of k_{dj}^s as a function of T_g at two different E/N values. The small increase of k_{dj}^s with decreasing gas temperature should be compared with the strong dependence of k_d^s on T_g. This behaviour is due to the fact that vibrational levels responsible for dissociation in JVE are those far from the v-t deactivation region (i.e. far from the Boltzmann tail) while, in PVM, only levels belonging to the tail can dissociate. The strong dependence of k_d^s on T_g is therefore attributed to the sensitivity of the Boltzmann tail of the N_v distribution to the gas temperature. In contrast, the apparent independence of k_{dj}^s on T_g is due to the small dependence of the N_v portion far from the tail on T_g. It is worth noting that the distribution of vibrational levels in both JVE and PVM are practically the same.

A comparison between k_{dj}^s, k_d^s and k_d^e $(v = 0) n_e$ (s^{-1}) is shown in Fig. 7. One appreciates that k_{dj}^s values are always larger than the corresponding k_d^e $(v = 0) n_e$ values, the deviation being one order of magnitude at E/N = 2×10^{-16} V cm^2 and $T_g \leqslant 350$ K where k_d^s also overcomes k_d^e $(v = 0) n_e$.

Figure 9 and 11 show k_{dj}^s, k_d^s and k_d^e $(v = 0) n_e$ as a function of p and of n_e respectively. The dependence of k_{dj}^s on this parameters is small when compared with the corresponding variation of k_d^s values. Once again this point can be understood on the basis of the different portions of the N_v distributions participating in JVE and PVM.

The behaviour of N_v distributions at different pressures is very similar in the two mechanisms (see Fig. 10). The dramatic dependence of k_d^s on v-v and v-t rates has been pointed in previous work[4-6]. The accuracy of k_{dj}^s values depends on e-v and DEM rate coefficients rather than on v-v and v-t ones. This point can be understood by noting that the portion of the N_v distribution involved in JVE (i.e., the first levels and the plateau) does not depend too much on v-v and v-t cross sections. For example an increase by a factor of 2 of all v-v rates and a decrease by the same factor of corresponding v-t rates produce no appreciable variations in k_{dj}^s values ($n_e = 10^{11}$ cm^{-3}, $T_g = 500$ K, p = 6,66 mbar), while the same variations cause differences up to two orders of magnitude in k_d^s values. On the other hand the accuracy of the present k_{dj}^s values strongly depends on both DEM and e-v rates. In turn k_d^e (v) rates depend on the relevant cross sections as well as on the electron energy distribution function.

2.3.3 Recombination Process

Different approaches can be used for introducing the recombination process as well as the influence of atoms on both PVM and JVE. The simplest one is an extension of the method utilized in Ref.[11] for the dissociation of H_2 under thermal conditions ($n_e = 0$).

It consists of reversing the processes (XII) and (XIII). One can therefore write the following equations for $N_{v'+1}$, $N_{v'}$ and N_v:

$$dN_{v'+1}/dt = - (P^{H_2}_{v'+1,v'}N_{H_2} + P^H_{v'+1,v'}N_H) N_H^2 + (P^{H_2}_{v',v'+1}N_{H_2} + P^H_{v',v'+1}N_H)N_{v'} +$$

$$+ \sum_{i=0}^{v'} (P^{v',v'+1}_{i,i-1} N_{v'}N_i - P^{v'+1,v'}_{i-1,i} N_H^2 N_{i-1} + n_e \sum_{i=0}^{v'} k_d^e (i) N_i \qquad (7)$$

$$dN_{v'}/dt = - [(P^{H_2}_{v',v'-1}N_{H_2} + P^H_{v',v'-1}N_H) + (P^{H_2}_{v',v'+1} N_{H_2} + P^H_{v',v'+1}N_H)] N_{v'} +$$

$$+ (P^{H_2}_{v'-1,v'}N_{H_2} + P^H_{v'-1,v'}N_H)N_{v'-1} + (P^{H_2}_{v'+1,v'}N_{H_2} + P^H_{v'+1,v'}N_H)N_H^2 +$$

$$+ \sum_{i=0}^{v'} [P^{i,i+1}_{v'+1,v'} N_H^2 + P^{i,i-1}_{v'-1,v'} N_{v'-1} - (P^{i,i+1}_{v',v'-1} + P^{i,i-1}_{v',v'+1})N_{v'}]N_i +$$

$$+ (P^{v'+1,v'}_{v'-1,v'} N_{v'-1} - P^{v',v'+1}_{v'+1,v'} N_{v'})N_H^2 +$$

$$+ n_e \sum_{i=0}^{v'+1} (k^e_{i,v'}N_i - k^e_{v',i}N_{v'}) - n_e k_d^e(v')N_{v'} \qquad (8)$$

$$dN_v/dt = - [(P^{H_2}_{v,v-1} N_{H_2} + P^H_{v,v-1}N_H) + (P^{H_2}_{v,v-1} N_{H_2} + P^H_{v,v+1}N_H)]N_v +$$

$$+ (P^{H_2}_{v-1,v} N_{H_2} + P^H_{v-1,v}N_H)N_{v-1} + (P^{H_2}_{v+1,v} N_{H_2} + P^H_{v+1,v}N_H)N_{v+1} +$$

$$+ \sum_{i=0}^{v'} [P^{i,i+1}_{v+1,v}N_{v+1} + P^{i,i-1}_{v-1,v} N_{v-1} - (P^{i,i+1}_{v,v-1} + P^{i,i-1}_{v,v+1})N_v]N_i +$$

$$+ (P^{v'+1,v'}_{v-1,v} N_{v-1} - P^{v'+1,v'}_{v,v+1} N_v) N_H^2 +$$

$$+ n_e \sum_{i=0}^{v'+1} (k^e_{i,v}N_i - k^e_{v,i}N_v) - n_e N_v k_d^e (v) \qquad (9)$$

where

$$P_{v'+1,v'} = P_{v',v'+1} [\exp - (E_{v'}/kT_g)]/K_{eq} Q_v \qquad (10a)$$

$$P^{v'+1,v'}_{v,v+1} = P^{v',v'+1}_{v+1,v} [\exp - (E_{v+1} + E_{v'} - E_v)/kT_g]/K_{eq}Q_v \qquad (10b)$$

$$N_H = 2N_{v'+1}; \quad N_{H_2} = N^0_{H_2} - N_{v'+1} \qquad (10c)$$

Equation (7) disregards the possibility of atom recombination in the presence of electrons (i.e. the reverse of processes XIV) while the influence of atoms on the vibrational relaxation comes from terms $P^H_{v,v+1} N_H N_v$ and so on in the different equations. Eqs. (10a) and (10b) are a direct consequence of the detailed balance applied to Eqs. (XII)–(XIII), while K_{eq} and Q_v represent the equilibrium constant of the process $H_2 \rightleftarrows 2H$ and the vibrational partition function of H_2 respectively. These values have been taken from[12].

The v-t rate coefficients relative to

$$H + H_2(v) \xrightarrow{\;P_{v,v-1}^H\;} H + H_2(v-1) \tag{XV}$$

are strongly affected by chemical effects (exchange and reaction). The experimental value of Ref.[13] has been used for the 1-0 transition ($P_{10}^H = (3 \pm 1.9) \, 10^{-13} \; cm^3 \, s^{-1}$, $T_g = 300$ K. The $v \neq 1$ rates have been estimated in Ref.[9].

The system of vibrational master equations has been numerically integrated with the same initial condition described in Sect. 2.3.1. Figure 15 reports the N_v distribution obtained in JVE with allowance for recombination at different times. One can note that at a time of 10^{-3} s the N_v distribution with recombination is very similar to that calculated without recombination. The only part affected by recombination and by the presence of atoms is the tail of N_v-distribution, which is strongly over-estimated in JVE without recombination (see Fig. 8). This is due to the chemical deactivation, which increases the depopulation of higher vibrational levels. These considerations suggest that the k_d^s values (which depend on the tail of N_v distribution) are smaller than the corresponding values calculated without atoms.

On the other hand k_{dj}^s values, which do not depend on the tail of N_v distributions, are less sensitive to the presence of atoms.

It should be noted that the quasi-stationary values reported in Figs. 7–11 are reached in times of the order of 10^{-3} s. After $t = 10^{-3}$ s the increased atom concentration deactivates the vibrationally excited molecules, since the deactivation rate of process XV can exceed the pumping e-v rate $n_e k_{01}^e$. In this case one observes a strong decrease of the vibrational temperature θ_1, with dramatic consequence on the vibrational kinetics. As a result at $t = 10^{-2} - 10^{-1}$ s, one observes the disappearance of the N_v plateau resulting from the v-v pumping processes. The plateaus appearing in Fig. 15 ($t = 10^{-2}$, 10^{-1} s) are in this case due to the recombination process, which populates levels near the continuum (see Sect. 3.2.1).

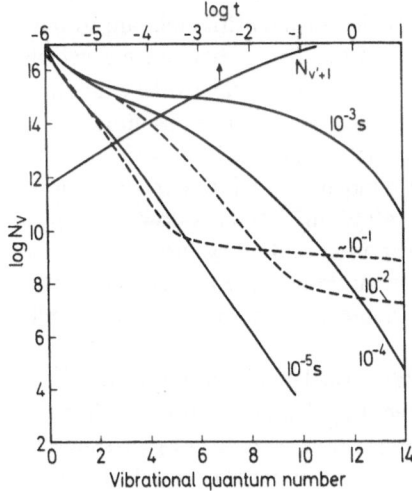

Fig. 15. Vibrational populations at different times and half the atom concentration as a function of time for JVE with recombination ($n_e = 10^{12}$ cm^{-3}, $N_{H_2}^0 \sim 10^{17}$ cm^{-3}, $E/N = 3 \times 10^{-16}$ V cm^2, $T_g = 300$ K). (From Ref.[9])

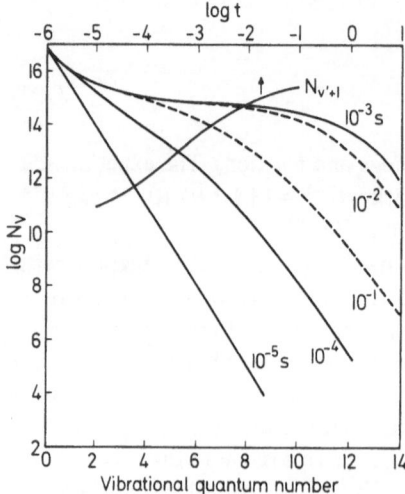

Fig. 16. Vibrational populations at different times and half the atom concentration as a function of time for JVE with recombination from Eqa. 15–19 (same conditions of Fig. 15; $E/N = 2 \times 10^{-16} \, V \, cm^2$) (From Ref.[9])

Figure 16 shows the JVE N_v distribution with recombination at $E/N = 2 \times 10^{-16} \, V cm^2$. In this case the results look very similar to those reported for JVE and PVM without recombination, since the density of H atoms does not exceed 5%.

Only at $t = 10^{-1}$ s deactivation by H atoms becomes important in this case.

2.3.4 Experimental Data

The existence of vibrational disequilibrium in a hydrogen flow discharge ($p - 20 \, mbar$, $E/N - 3 \times 10^{-16} \, V cm^2$) has recently been ascertained in direct measurements of the vibrational temperature of the outflowing gas by means of a stimulated Raman scattering technique[14]. ($T_v = 2300 \, K$, $T_g = 600 \, K$, at reactor's exit).

Measurements of the extent of H_2 dissociation in flowing discharges operated at pressures up to a few mbar are available and have been analyzed in Ref.[15]. Uncertainities in the selection of rate constants and of electric quantities relevant to the discharge model are rather high. Rate constants for the dissociation process, calculated for DEM from $v = 0$ only (k_d^e) and utilizing a maxwellian edf, can fit the experimental data. The accuracy of this fit is likely to be within one order of magnitude.

Measurements in a higher pressure range (6–55 mbar) have been performed in our laboratories[16]. A tubular flow reactor surrounded by a calorimetric water jacket was operated at 35 MHz, power densities of 4–40 W cm^{-3} and gas flow rates up to 0.05 mol s^{-1}. Values of the estimated gas temperatures plotted against power density have been reported in Fig. 17 for various conditions. Power densities are higher than those utilized in Ref.[14] quoted above and one appreciates that values of T_g are all below $T_v = 2300 \, K$.

In Fig. 18 the ratio between the rate constant calculated according to Ref.[15], and the corresponding experimental rate constant k_{exp} has been plotted against power density for various values of pressure and gas flow rates. This figure shows a gradual transition from a situation in which DEM predominated (low p, high $\bar{W}$) to

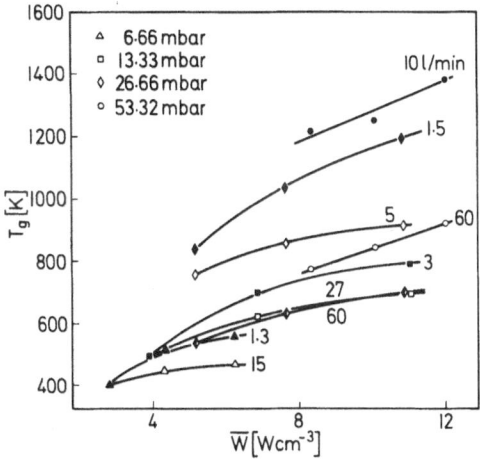

Fig. 17. Calculated values of the kinetic gas temperatures, T_g as a function of the power density $\overline{W}$ at different pressures and gas flow rates (From Ref.[16])

a situation where the contribution of DEM is reduced almost to zero (high p, high flow rates, low $\overline{W}$). Rate constants calculated according to JVE (k_{dj}^s) have been compared with k_d^e and the experimental values k_{exp} in Table 1 from Ref.[9] for different experimental conditions. Calculated θ_1 have also been included and appear to be reasonable.

It should be noted that dissociation cross sections used in JVE refer to excitation to the b $^3\Sigma_u$ state only. Actual dissociation cross sections include excitation to all triplet H_2 states. Use of experimental cross sections and of edf's which take into account the simultaneous presence of vibrational disequilibrium and of atoms yield values of k_{dj}^s, also reported in Table 1, which are in better agreement with k_{exp}. Further details can be found in Ref.[2].

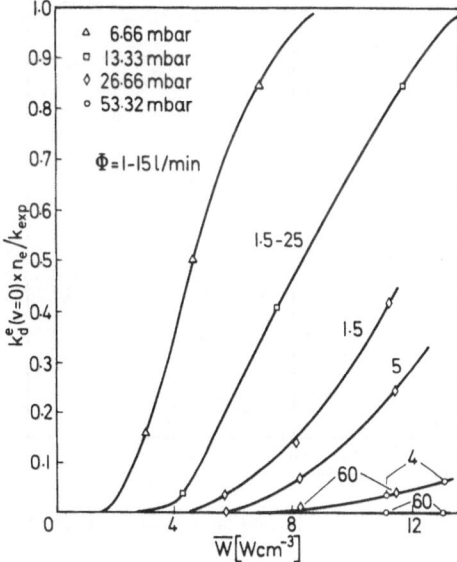

Fig. 18. $k_d^e (v = 0) n_e/k_{exp}$ ratio as a function of power density, at different pressures and gas flow rates (From Ref.[16])

Table 1. A comparison of theoretical and experimental[16] dissociation constants

E/N V cm^2	T_g (K)	p (mbar)	n_e cm^{-3}	k_{dj}^s [b] s^{-1}	k_{dj}^s [c] s^{-1}	k_{exp} s^{-1}	θ_1[d] K
4.1(−16)[a]	700	26.7	3.1(11)	8.4	36.8	114	2840
5.8(−16)	910	26.7	1.9(11)	20.0	75.0	149	2750

[a] $5.8(-16) = 5.8 \cdot 10^{-16}$.
[b] Cross sections and edf of Ref.[9].
[c] Cross sections of Corrigan (J. Chem. Phys. *43*, 4381, 1965) and edf of Ref.[2].
[d] θ_1 is the vibrational temperature from Ref.[9].

3 The Dissociation of Other Diatomic Molecules

3.1 Nitrogen[6a, 6b]

In molecular nitrogen large *e-v* and *v-v* rates are associated with unusually low *v-t* rates. This represents a "crossing" of conditions favourable to the PVM, as discussed in 2.3.1. This situation is immediately reflected in the N_v distributions of Fig. 19. These distributions consist of a Treanor distribution followed by a long plateau extending up to the dissociation limit. The Boltzmann tail has practically disappeared, as consequence of the low *v-t* rates, and the influence of T_g on the distribution is limited. The behaviour of the N_v distributions is reflected on the k_d^s values reported

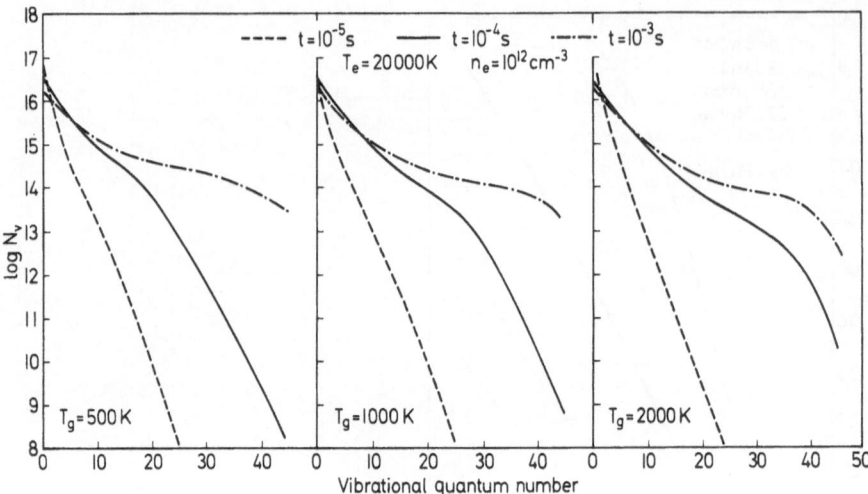

Fig. 19. Population densities (log N_v) as a function of vibrational quantum number at different times in molecular nitrogen (From Ref.[6b])

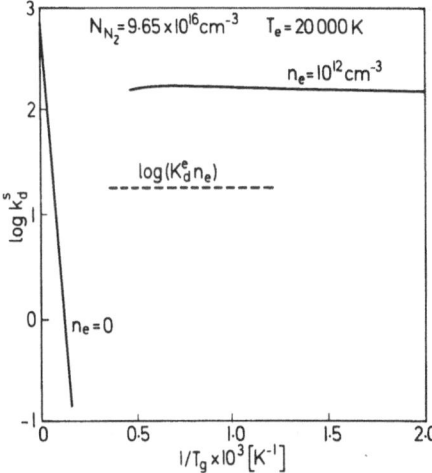

Fig. 20. Dissociation constants log k_d^s in molecular nitrogen as a function of gas temperature (From Ref.[6b])

in Figs. 20 and 21. The dependence on T_g and on p is strongly reduced with respect to H_2 and large values of the dissociation rate constants are obtained which can be as much as two orders of magnitude higher than the corresponding DEM rate constants $k_d^e (v = 0) n_e$ (Fig. 21). N_v distributions and k_d^s values are also expected to be very sensitive to the presence of species which enhance the rate of v-t processes, e.g. H_2[6a]. This is clearly brought out in the N_v distributions of Fig. 22b where a Boltzmann tail is again present with its depressing influence on the PVM rate constant k_d^s (Fig. 21).

A depressing action is also expected on the part of N-atoms. N_2 (v)-N v-t rates are not available but can be estimated on the basis of the known N_2 (v)-O v-t rates[17].

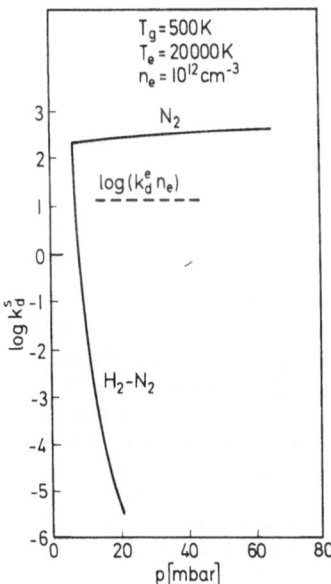

Fig. 21. Dissociation constants log k_d^s as a function of pressure in molecular nitrogen (PVM) and for H_2–N_2 mixtures. In the last case p_{N_2} = 6.66 mbar (From Ref.[6a])

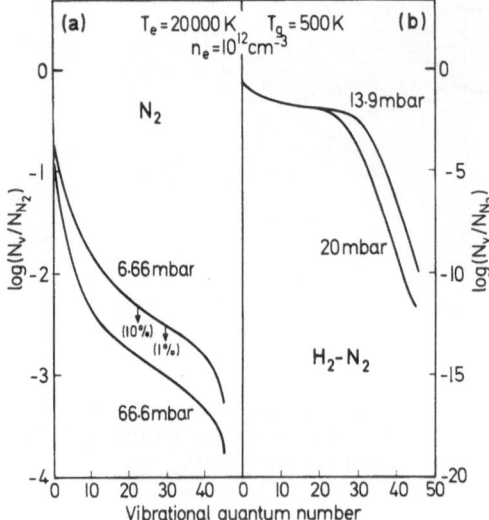

Fig. 22a and b. Population densities as a function of vibrational quantum number at different pressures in (a) pure nitrogen and (b) in H_2-N_2 mixtures (p_{N_2} = 6.66 mbar). The arrows at p = 6.66 mbar indicate the estimated v's for the onset of the Boltzmann tails due to the presence of 1 and 10% of nitrogen atoms)

In Fig. 22a the arrows mark the onset of a Boltzmann tail for 1% and 10% of atomic nitrogen. The corresponding v's have been calculated according to Ref.[8]. The occurrence of non-Boltzmann distributions of the type illustrated in Figs. 19 and 22 has been demonstrated experimentally.

Spectroscopic observations of the emission of the first positive system of N_2 ($B^3\Pi_g - A^3\Sigma_u^+$) in electrical discharges[18] have been interpreted by assuming that the $B^3\Pi_g$ state is populated by electron excitation from the ground $X^1\Sigma_g^+$ state; the distribution of vibrational populations observed in the B state is consistent with the

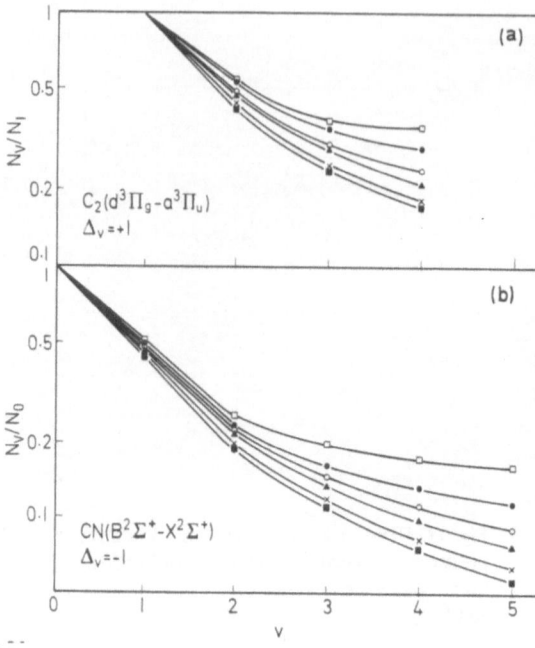

Fig. 23a and b. Relative population of the various levels of a C_2 and b CN as a function of vibrational quantum number at different pressures (From Ref.[19]). (□: p = 6.66 mbar; ●: p = 13.3; ○: p = 20; ▲: p p = 26.7; x: p = 33.3; ■: p = 40 mbar)

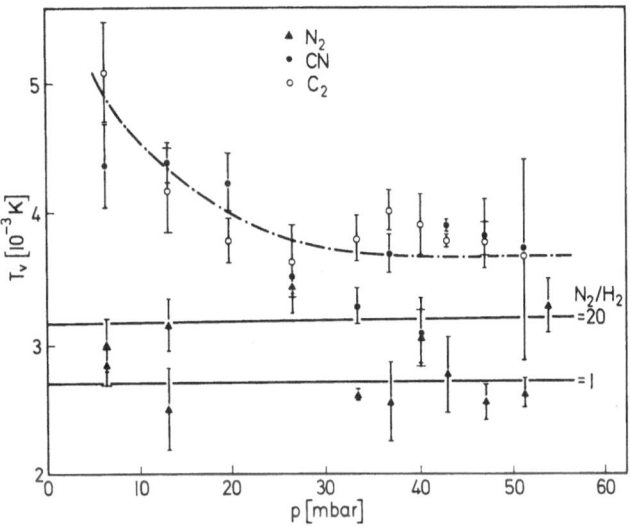

Fig. 24. Vibrational temperatures T_v of $N_2(C^3\Pi_u)$, CN $(B^2\Sigma^+)$ and $C_2(d^3\Pi_g)$ states as a function of pressure (From Ref.[19])

distribution of vibrational levels in the X state similar to those reported in Figs. 19 and 22. Figures 23–25 illustrate results obtained in our laboratories[19] with $N_2/H_2/CH_4$ mixtures flowing in rf discharges operated under conditions similar to those described for hydrogen in Sect. 2.4. The presence of non-Boltzmann distributions of the vibrational levels in the $C_2(d^3\Pi_g)$ and the $CN(B^2\Sigma^+)$ states is apparent from Fig. 23 for $v > 2$. Vibrational temperatures, derived from the first three levels, decrease with increasing pressure for C_2 and CN and are independent of pressure for $N_2(C^3\Pi_u)$ (Fig. 24). Rotational states obey Boltzmann law up to high rotational quantum numbers[19, 20]. A rapid increase with pressure of the rotational temperature T_r up to 4300 K is apparent from Fig. 25 for the three emitting species.

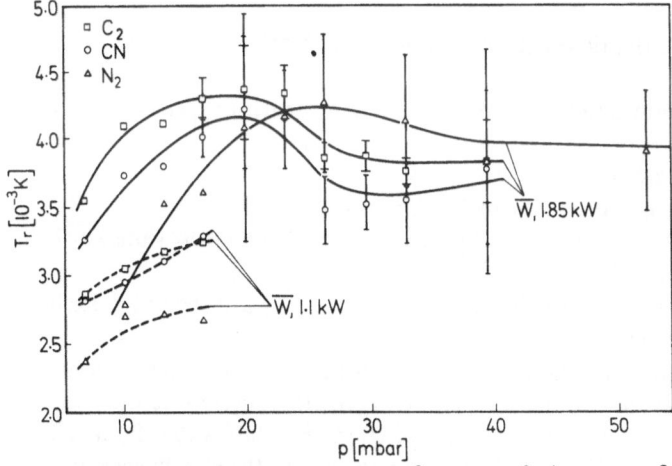

Fig. 25. Rotational temperatures T_r of $N_2(C^3\Pi_u)$, $CN(B^2\Sigma^+)$ and $C_2(d^3\Pi_g)$ states as a function of pressure (From Ref.[19])

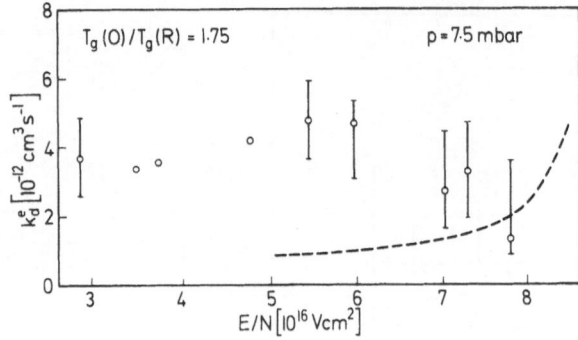

Fig. 26. Experimental values of dissociation constants k_d^e cm^3 s as a function of E/N in pure nitrogen (– – – theoretical curve from DEM; see Ref.[22a] for details). $T_g(O)$ and $T_g(R)$ is the gas temperature at the center and at the boundary of the discharge respectively

Direct determination of non-Boltzmann distributions of the vibrational levels of the ground $N_2(X^1\Sigma_g^+)$ state has recently been performed by a new diagnostic technique, coherent anti-Stokes Raman spectroscopy[21] with results consistent with calculated distributions. Kinetic data on N_2 dissociation amenable to a comparison with theoretical predictions are scanty. Data from Ref.[22a] can however be quoted and are reported in Fig. 26. The dashed line has been calculated on the assumption that dissociation takes place via predissociated electronic states excited by direct electron impact. Observed dissociation rates are higher and a much better agreement has been claimed with dissociation rates calculated on the basis of a pure vibrational mechanism[22a, 22b].

3.2 Oxygen

3.2.1 Vibrational Distributions and Dissociation Rates[5, 23]

The characteristic features to be considered for the molecular oxygen system are the following
1) e-v pumping rates are small.
2) v-v rates overcome v-t rates in the molecular system.
3) Process $O_2(v) + O \longrightarrow O_2(v-1) + O$ is very effective in removing vibrational energy by v-t transfer[24, 25].
4) Dissociation via direct electron impact to both the $A^3\Sigma_u^+$ and the $B^3\Sigma_u^-$ excited electronic states is very effective.

These features have important implications for the dissociation mechanisms.

The presence of oxygen atoms is crucial. Oxygen atoms deactivate the first vibrational level of the ground electronic state of O_2, thus reducing the vibrational temperature $\theta_1 = E_{10}/k\ln(N_0/N_1)$ which is responsible for the effectiveness of the vibrational mechanism. An estimate of the maximum concentration of $v = 1$ level (and therefore of θ_1) in JVE can be made by assuming the following processes to be responsible for the production and destruction of O_2 ($v = 1$) species

$$e + O_2\ (v = 0) \xrightarrow{\ k^e_{01}\ } e + O_2\ (v = 1) \qquad \text{(XVI)}$$

$$O_2\ (v = 1) + O \xrightarrow{\ P^O_{10}\ } O_2\ (v = 0) + O \qquad \text{(XVII)}$$

$$e + O_2\ (v = 0) \xrightarrow{\ k^e_d(v=0)\ } e + 2O \qquad \text{(XVIII)}$$

One obtains[23] a maximum value of N_1 at a time

$$t_{max} = \left(2 N_{O_2(v=0)} k^e_d\ P^O_{10}\ n_e\right)^{-\frac{1}{2}} \qquad \text{(11)}$$

$$(N_1)_{max} \approx n_e\ k^e_{01}\ N_{O_2}(v = 0)\ t_{max} \qquad \text{(12)}$$

The quenching of N_1 (and therefore of θ_1) is practically unaffected by the presence of oxygen atoms for $t < t_{max}$ the reverse being true for $t > t_{max}$.

Equations (11)–(12) indicates that appreciable values of N_1 (θ_1) can be obtained for discharge conditions in which large values of the pumping rate k^e_{01} combine with small values of the dissociation constant k^e_d. These favourable conditions hold for E/N values less than 10^{-16} V cm^2, while for E/N $> 10^{-16}$ one expects very small vibrational temperatures (i.e. $\theta_1 \sim T_g$) with important consequences on the vibrational kinetics.

An aspect which should also be duly considered is the influence of the recombination process on the N_v distributions. In fact, when oxygen atoms reach a large concentration one expects that levels lying near the continuum will be affected by the recombination process. A simplified picture of the situation can be drawn by assuming that

(i) DEM for $v = 0$ is the only atom producing mechanism; (ii) the levels near the continuum are dominated by v-t exchanges (v-v process neglected); (iii) a region of levels $v^* < v \leqslant v'$ is populated by a cascade model. Points (i) (ii) (iii) can be simulated as

$$e + O_2(v = 0) \xrightarrow{\ k^e_d\ } e + 2O \qquad \text{(XIX)}$$

$$O + O + M \longrightarrow O_2(v' + 1) + M \xrightarrow{\ P^M_{v'+1,v'}\ } O_2(v') + M \qquad \text{(XX)}$$

$$O_2(v') + M \xrightarrow{\ P^M_{v',v'-1}\ } O_2(v'-1) + M \qquad \text{(XXI)}$$

$$O_2(v) + M \xrightarrow{\ P^M_{v,v-1}\ } O_2(v-1) + M \qquad \text{(XXII)}$$

From this simplified kinetics scheme, one obtains[23]

$$N_{v'} = \frac{P^M_{v'+1,v'}}{P^M_{v',v'-1}} \ N^2_O \tag{13}$$

$$N_v = N_{v+1} \frac{P^M_{v+1,v}}{P^M_{v,v-1}} = \frac{P_{v'+1,v'}}{P^M_{v,v-1}} \ N^2_O \tag{14}$$

Equations (13)–(14) show that a portion of levels near v' increases its concentration with a N^2_O law, while the ratio $\dfrac{N_v}{N_{v+1}}$ is determined by the deactivation rates of $(v+1)$ and (v) levels.

Another interesting aspect is the possibility of obtaining large concentration of level v' under special conditions (i.e., large $P^M_{v'+1,v'}$ and small $P^M_{v',v'-1}$ rates). These high concentration of v' species can redissociate according to

$$O_2(v') + M \longrightarrow O_2(v'+1) + M \longrightarrow 2O + M \tag{XXIII}$$

i.e., a recombination assisted dissociation process.

Features of N_v distributions and rates in oxygen, will be illustrated by selected numerical results obtained in different E/N regions.
1) $E/N > 10^{-16}$ V cm^2 (JVE + Recombination)

Conditions: $- E/N = 1 \cdot 5 \ \ 10^{-16}$ V cm^2, $n_e = 10^{11}$ cm^{-3}, p = 26.7 mbar
$T_g = 500$ K.

Values of the population densities of levels $v = 1, 2$ and half the concentration of O-atoms have been plotted against time in Fig. 27. Deactivation by O-atoms causes populations to decrease for $t > t_{max}$ bringing N_1 to it's Boltzmann value at T_g.

The maximum vibrational temperature θ_1 (t_{max}) is $\simeq 700$ K and close to $T_g = 500$ K. The concentration of oxygen atoms at $t = t_{max}$ is approximately 10^{15} cm^{-3}, i.e., 1% of the total concentration. In view of the small vibrational non-equilibrium, the v-v pumping mechanism is of minor importance.

Figure 28 shows the N_v distribution at $t = t_{max}$ and for times $t > t_{max}$. In Fig. 28a we have also reported the instantaneous Treanor's distributions at the relevant $\theta_1(t)$'s values. These distributions, which are analytical solutions of the system of vibrational master equations including v-v rates only, should represent upper limits to the actual N_v distributions, which include all v-t deactivating processes.

One appreciates that the low lying vibrational levels only satisfay Treanor's distributions, while from approximately $v = 10$ on a plateau is present in the N_v distributions, the population of which is much higher than the corresponding Treanor's values. Concentrations belonging to the plateau grow with a N^2_O law, while the N_v/N_{v+1} ratio approximately fits Eq. (14). This plateau should therefore be attributed to the recombination process, as outlined above.

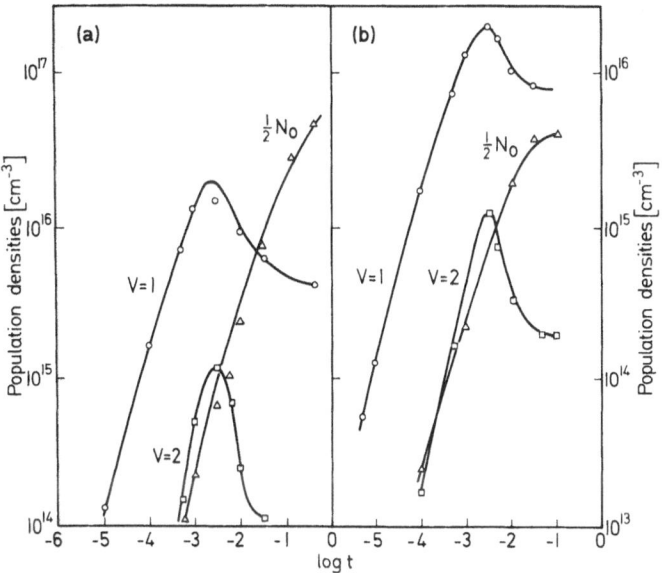

Fig. 27a and b. Population densities of $v = 1$ and $v = 2$ levels and half the atom concentration as a function of time ($n_e = 10^{11}$ cm^{-3}, p = 26.7 mbar, T_g = 500 K, E/N = 1.5 × 10^{-16} V cm^2); a θ' = 1.04 × 10^6 K; b θ' = 6.5 × 10^6 K (From Ref.[23])

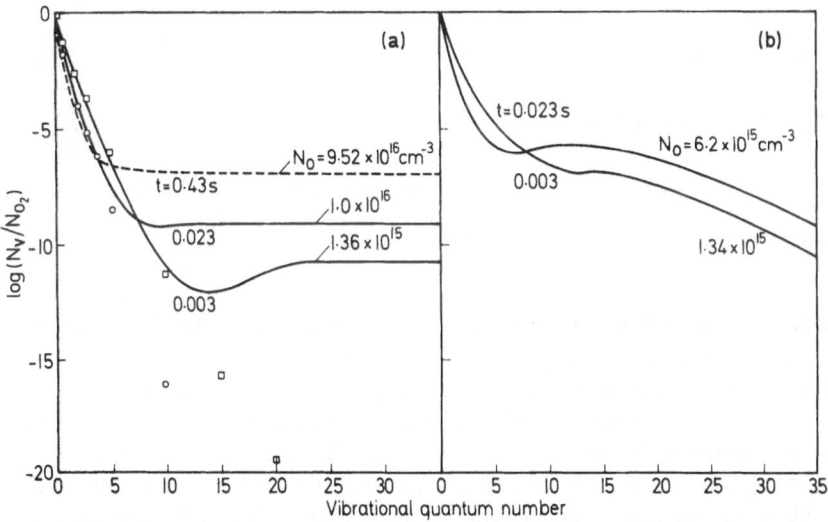

Fig. 28a and b. Values of the population densities log N_v/N_{O_2} as a function of vibrational quantum number (same notations as in Fig. 27); a ○ □ Treanor's distributions at t = 2.3 × 10^{-2} s and t = 3 × 10^{-3} s (From Ref.[23])

Table 2. Normalized first order dissociation constants (s^{-1}) for processes XXIV, XXV, XXVI (see text) $(E/N = 1.5 \times 10^{-16} \, V \, cm^2$, $n_e = 10^{11} \, cm^{-3}$, $p = 26.7$ mbar, $T_g = 500$ K)

$\theta' = 5.6 \times 10^6 \, K^a$

time (ms)	k(XXIV)	k(XXV)	k(XXVI)
10	0.56	9. 10^{-6}	4.5 10^{-2}
20	0.56	3.4 10^{-5}	1.1 10^{-1}
30	0.56	5. 10^{-5}	1.4 10^{-1}
50	0.56	6.2 10^{-5}	1.6 10^{-1}
90	0.56	6.2 10^{-5}	1.6 10^{-1}

$\theta' = 1.04 \times 10^6 \, K$

133	0.56	1.44 10^{-2}	7.7 10^{-4}
233	0.56	4.9 10^{-2}	1.7 10^{-3}
333	0.56	7.4 10^{-2}	2.2 10^{-3}
433	0.56	8.6 10^{-2}	2.5 10^{-3}

a   The quantity θ' enters in the adiabaticity factor of SSH theory used in this work for calculating v-t and v-v rates (see Ref.[4] and [71]).

Table 2 compares the pseudo first order dissociation constant for the process

$$e + O_2(v = 0) \longrightarrow e + 2O \qquad \qquad \text{(XXIV)}$$

$$O + O_2(v') \longrightarrow O + 2O \qquad \qquad \text{(XXV)}$$

$$O_2 + O_2(v') \longrightarrow O_2 + 2O \qquad \qquad \text{(XXVI)}$$

These constants have been normalized to the O_2 $(v = 0)$ concentration i.e., we compare $k_d^e \, (v = 0) \, n_e$, $N_{v'} P^O_{v',v'+1} \dfrac{N_O}{N_{O_2(v=0)}}$, $N_{v'} P^{O_2}_{v',v'+1} \dfrac{N_{O_2}}{N_{O_2(v=0)}}$.

It should be noted that the contribution coming from reaction XXVI[23] is approximately 30% of the corresponding k_d^e $(v = 0) \, n_e$ for $\theta' = 5.5 \times 10^6$ K, while it becomes insignificant for $\theta' = 1.04 \times 10^6$ K. If one increases the electron density to $n_e = 10^{13} \, cm^{-3}$, other conditions being equal, a consistent vibrational nonequilibrium with respect to the gas temperature can now be established.

At $t = t_{max}$ the vibrational temperature θ_1 reaches a value of 1650 K, the system being able to sustain appreciable vibrational populations of higher levels by means of v-v exchanges. The N_v distribution (see Fig. 29) is, at $t = t_{max}$, very close to the corresponding Treanor's distribution which in this case overestimates the populations. As time evolves, the concentration of deactivating oxygen atoms increases and the

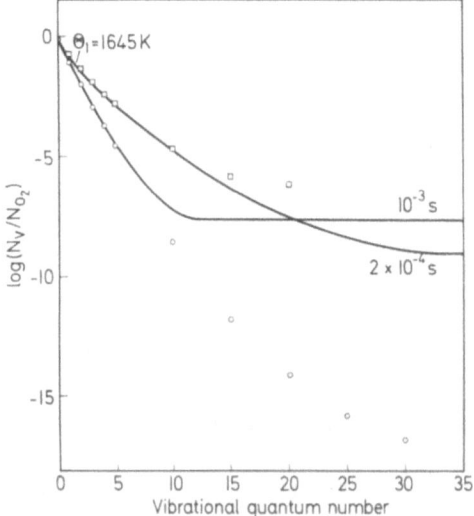

Fig. 29. Values of the population densities ($\log N_v/N_{O_2}$) as a function of vibrational quantum number. $\circ$ $\square$ Treanor's distributions at $t = 10^{-3}$ and 2×10^{-4} s; $\theta' = 1.04 \times 10^6$ K, $n_e = 10^{13}$ cm^{-3}, p = 26.7 mbar, $T_g = 500$ K, E/N = 1.5×10^{-16}Vcm^2 (From Ref.[23])

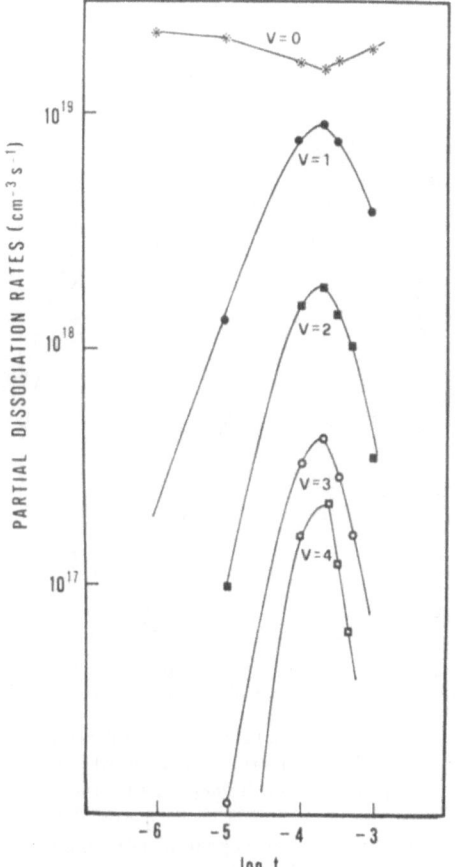

Fig. 30. Values of the partial dissociation rate ($n_e N_v k_d^e(v)$) cm^{-3} s^{-1}) as a function of time for $v = 0-4$ vibrational levels (same notations as in Fig. 29) (From Ref.[23])

vibrational temperature falls toward the translational one. At $t = 10^{-3}$ s θ_1 is equal to 950 K and the corresponding Treanor's distribution holds now only for the first few vibrational levels ($v < 10$), while for $v > 10$ the plateau generated by the recombination process strongly overcomes the Treanor's distributions.

Figure 30 compares the partial dissociation rates coming from the vibrational levels $v \neq 0$ with the corresponding $v = 0$ case $\left[\text{i.e., } e + O_2(v) \xrightarrow{k_d^e(v)} e + 2O; \text{partial rate} = n_e N_v k_d^e(v) \right]$.

One can note that at $t = t_{max}$ the contribution coming from $v = 1$ is of the same order of magnitude as the $v = 0$ one, decreasing its importance for $t > t_{max}$.

2) $E/N < 10^{-16}$ V cm^2. (PVM + recombination)
Conditions: $10^{-17} < E/N < 10^{-16} \sim KT_e$ 5000 K, $p = 6.66$ mbar, $T_g = 500$ K

Figure 31 shows the temporal evolution of the population of first few vibrational levels as well as the production of oxygen atoms.

As a consequence of the small dissociation rate, the first vibrational levels reach quasi-stationary values in times of the order of $(n_e k_{10}^e)^{-1}$. At the vibrational temperature $\theta_1^s \sim 2270$ K v-v processes will transfer the introduced quanta up to the dissociation level. PVM therefore prevails under these conditions. The production of oxygen atoms is, however, very small (Fig. 31).

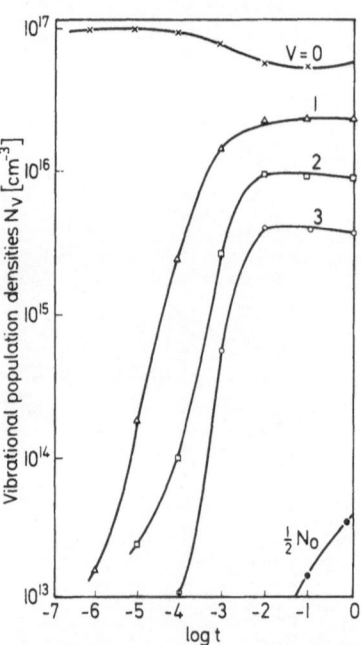

Fig. 31. Vibrational population densities and half the atom concentration as a function of time for the pure vibrational mechanism ($n_e = 10^{12}$ cm^{-3}, $E/N < 10^{-16}$ V cm^2, $p = 6.66$ mbar, $T_g = 500$ K, $\theta' = 2.6 \times 10^6$ K) (From Ref.[23])

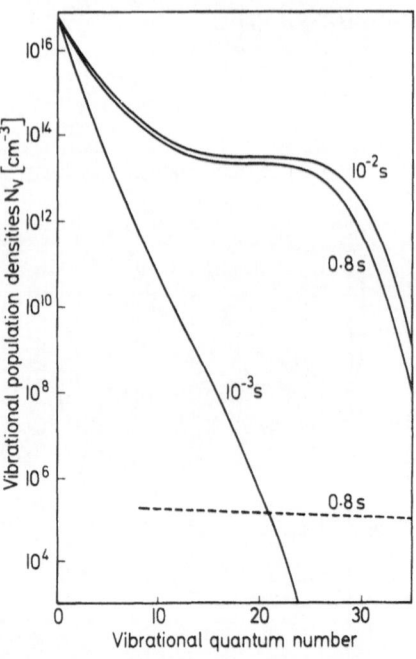

Fig. 32. Vibrational populations densities (cm^{-3}) as a function of vibrational quantum number at different times for the same conditions of Fig. 31 --- an estimation of the "recombination plateau" at $t = 0.8$ s (From Ref.[23])

Figure 32 shows the complete N_v distribution at different times. Three portions of the N_v distribution are present, the Treanor, the plateau and the Boltzmann tail. The plateau is now controlled by v-v exchanges because the contribution coming from the recombination process is negligible as a consequence of the low concentration of oxygen atoms. An estimated contribution of the recombination process is also shown in Fig. 32 at a time at which the oxygen atom concentration is practically stationary.

If the gas temperature is decreased to 300 K one obtains dissociation rates up to one order of magnitude larger than the corresponding rates at $T_g = 500$ K.

These rates, however, produce higher oxygen atom concentrations in the early part of the evolution, which, at once, self limit the dissociation rates.

3.2.2 Experimental Data

Dissociation of molecular oxygen in non-equilibrium plasmas has been subject of numerous experimental investigations, mostly at pressures below about 10 mbar[26-31].

Dissociation from electronic states excited by direct electron impact has generally been considered as the prevailing mechanism in this pressure range. The possibility of alternative mechanisms involving vibrationally excited ground state molecules has recently been explored at higher pressures (27 mbar)[32]. Apparatus and conditions are similar to those described for hydrogen. A method of titration of O-atoms with CO has been developed, appropriate to the higher pressure range investigated.

A reaction scheme which takes into account the most relevant processes commonly accepted to occur under plasma conditions is

$$O_2 + e \longrightarrow O + O + e \quad k^e \text{ (XXVII)} \qquad \text{(XXVII)}$$
$$O_2 + e \longrightarrow O_2(^1\Delta) + e \quad k^e \text{ (XXVIII)} \qquad \text{(XXVIII)}$$
$$O_2(^1\Delta) + e \longrightarrow O + O + e \quad k^e \text{ (XXIX)} \qquad \text{(XXIX)}$$
$$O + O_2 + O_2 \longrightarrow O_3 + O_2 \quad k^e \text{ (XXX)} \qquad \text{(XXX)}$$
$$O_2(^1\Delta) + O_3 \longrightarrow O + O_2 + O_2 \quad k^e \text{ (XXXI)} \qquad \text{(XXXI)}$$
$$O + O + O_2 \longrightarrow O_2 + O_2 \quad k^e \text{ (XXXII)} \qquad \text{(XXXII)}$$
$$O + O + O \longrightarrow O_2 + O \quad k^e \text{ (XXXIII)} \qquad \text{(XXXIII)}$$
$$O + O_3 \longrightarrow O_2 + O_2 \quad k^e \text{ (XXXIV)} \qquad \text{(XXXIV)}$$
$$O + \text{wall} \longrightarrow 1/2\,O_2 \quad k^e \text{ (XXXV)} \qquad \text{(XXXV)}$$

Net production of oxygen atoms by dissociative attachment

$$e + O_2 \longrightarrow O^- + O \qquad \text{(XXXVI)}$$

can be excluded in that the reverse reaction is very rapid.

The kinetic expression for atom production in the discharge zone is

$$\frac{dN_O}{dt} = 2[k^e(XXVII) + k^e(XXVIII)]N_{O_2} n_e - r_{rec} \tag{15}$$

which can be derived on the assumption of a stationary state for the concentrations of both O_3 and $O_2(^1\Delta)$, consistent with the negligible amounts of these species detected in the outflowing gas.

$k^e(XXVII)$ and $k^e(XXVIII)$ can be evaluated from the relevant cross sections $\sigma(u)$ and the discharge parameters. For reaction (XXVII), excitations leading to both the Herzberg system $(A^3\Sigma_u^+)$ and the Schumann Runge system $(A^3\Sigma_u^-)$, with threshold energies at 4.5 and 8.0 eV respectively, are considered as the most probable dissociative channels. For process (XXVIII) excitations to both the singlet $(a^1\Delta_g)$ and the singlet $(b^1\Sigma_g^+)$ state have been considered (threshold energies 0.96 and 1.63 eV respectively). Values of $k^e(XXVII)$ and $k^e(XXVIII)$ have been plotted against the reduced field in Fig. 33. The overall rate of O-atoms disappearance r_{rec} has been calculated in Ref.[32] by appropriate selection of literature values of the relevant rate constants ($k^e(XXX)$ and $k^e(XXXI)$ to $k^e(XXXV)$). In Fig. 34 the dependence of the extent of O_2 dissociation on residence time has been plotted as calculated with and without the inclusion of the dissociation channels via singlet states and by utilizing either maxwellian edf's or the more appropriate non-maxwellian edf of Ref.[33]. The experimental points, included in this figure, show quite clearly that the observed kinetics follows different patterns. This can also be appreciated from Fig. 35 where the degree of dissociation z_d has been plotted as a function of residence time NV/F° according to the equation

$$k_{exp}NV/F^\circ = - (z_d + 2\ln(1 - z_d)) \tag{16}$$

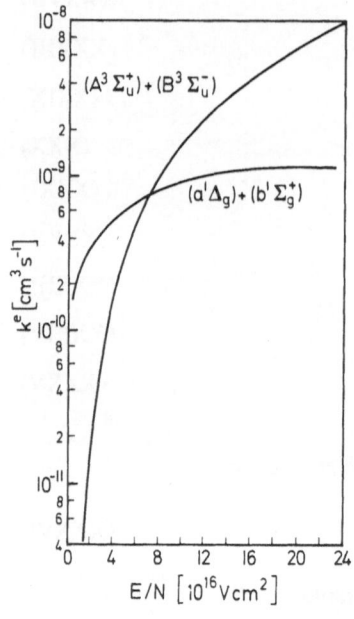

Fig. 33. Rate constants of direct electron excitation to both Σ and Δ singlet states and to both Schumann and Hertzberg systems as a function of E/N (From Ref.[32])

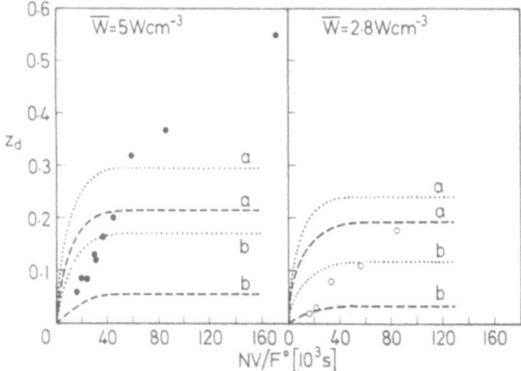

Fig. 34. Comparison of the experimental values of the degree of dissociation, z_d, with the ones calculated according to Eq. (15) as a function of residence time with the following notation: dotted and dashed lines refer to maxwellian and non maxwellian energy distribution functions, respectively, a and b, to the inclusion or exclusion respectively, of Δ singlets in the calculations (From Ref.[32])

(V (cm^3) = volume of discharge zone, F°(mol s^{-1}) = inlet molar flowrate of O_2) which represents the integrated form for a 1st order, plug-flow kinetics of dissociation, *without recombination* (i.e. r_{rec} = 0). The first order rate constant derived from these plots ($3-7$ s^{-1}) are in satisfactory agreement with values of k^e (XXVII) n_e calculated at the corresponding E/N values. Inclusion of dissociative channels via singlet states

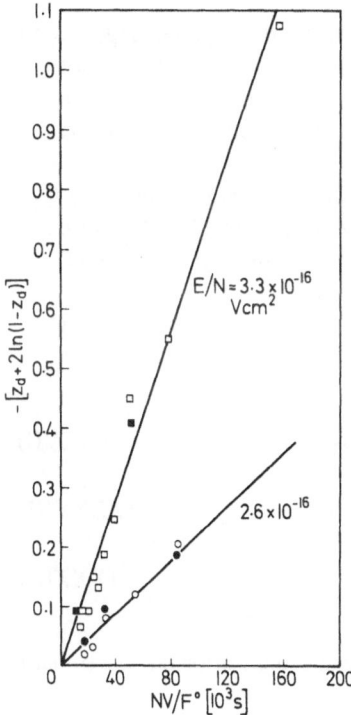

Fig. 35. First order, plug flow plots, of the experimental data at two values of E/N (From Ref.[32])

[k^e(XXVIII)n_e] would result in rate constants 5 to 10 times larger than the observed ones.

Attempts to fit the experimental data with the inclusion of recombination process, lead to values of the dissociation rate constant which increase with time by more than one order of magnitude.

The possible additional contribution of a "ladder climbing" vibrational mechanism to the rate of dissociation is however incompatible with the presence of a high concentration of oxygen atoms, as discussed in Sect. 3.2.1.

The recombination assisted dissociation process also discussed in Sect. 3.2.1 can contribute up to 30% to the rates calculated for a dissociative electron excitation mechanism. This calculated contribution is however insufficient to account for the present observations which correspond to almost complete suppression of recombination within the discharge region. In order to account for such a large suppression of recombination to the ground electronic state of O_2, the suggestion has been made[32] that the mechanism of recombination assisted dissociation, might be applied not only to the vibrational system of the ground electronic state, but, more advantageously, to the *long lived* metastables lying at 4.2–4.5 eV above ground state and close to the dissociation limit ($C^3\Delta_u$, $A^3\Sigma_u^+$, $C^1\Sigma_u^-$). This group of metastables should therefore be responsible for the observed suppression of recombination under the present experimental conditions. Under the experimental conditions of Ref.[32] the heterogenous recombination of the atoms to the ground state of O_2, occurring on the walls (reaction XXXV), can be shown to be small with respect to the rate of atom production. Similar considerations very likely apply also to the heterogenous losses of metastables, while their homogenous removal is replaced by various plasma processes.

The other fact brought out by the observed kinetics is the apparent suppression of the dissociation channel via Δ-singlets, which should make a rather large contribution to the process. A possible explanation is that removal of these metastables under discharge conditions effectively competes with their excitation to the dissociation limit.

In this connection, it should be mentioned that there is a series of resonant type cross-sections for reaction (XXXVI) in the energy range $0.9 - 2.0$ eV, which corresponds to vibrational quantum numbers $v = 5$–9 of the ground $O_2(X^3\Sigma_g^-)$ state, and which can be attributed to a $^1\Delta_g$ and $b^1\Sigma_g^+$ single states.

A possible route for the removal of singlet states could, therefore, be through the following process:

$$O_2(a^1\Delta_g, b^1\Sigma_g^+) + O_2 \longrightarrow O_2(X^3\Sigma_g^-)\,(v = 5\text{–}9) + O_2 \qquad\text{(XXXVII)}$$

$$O_2(v) + M \longrightarrow O_2(v - 1) + M \qquad\text{(XXXVIII)}$$

$$O_2(1) + O \longrightarrow O_2(v = 0) + O \qquad\text{(XXXIX)}$$

The main conclusions which can be derived from the analysis of the experimental results can be summarized as follows:

1) In this type of discharge both O_3 and O_2 $(a\,{}^1\Delta_g)$, are practically absent.

2) The main channels to dissociation are by electron excitation to the Schumann and Herzberg systems with practically no contribution of the channel via $O_2(a\,{}^1\Delta_g)$ and O_2 $(b\,{}^1\Sigma_g^+)$.

3) The use of maxwellian or of non maxwellian electron energy distribution functions is clearly reflected in the values of the rate constants.

4) When the contribution of O-atoms becomes appreciable there appear to be two opposing effects: depopulation of vibrational levels by the very effective $O_2(v) + O \longrightarrow O_2(v-1) + O$ $v - t$ transfer and repopulation of the higher levels by cascade from the recombination level. The effective suppression of O-atom recombination to ground state of O_2 within the discharge is probably linked to a recombination assisted dissociation involving the high lying metastables $(C\,{}^3\Delta_u, A\,{}^3\Sigma_u^+, C\,{}^1\Sigma_u^-)$.

3.3 Carbon Monoxide

Electrical gas discharges in carbon monoxide have received a great deal of attention due to their importance for the generation of laser beams in the infrared. Electron energy distribution functions have been calculated by different authors[34–36] (see in particular Ref.[34]). The influence of superelastic collisions on edf and related quantities has recently been treated[36] with results very similar to those discussed for hydrogen in Sect. 2. An increase of the vibrational temperature from $T_v = 300\,\mathrm{K}$ to $T_v = 3000\,\mathrm{K}$ increases the rate of ionization by one order of magnitude and causes a 20% increase of $\bar{u}_r$ at $E/N = 3 \times 10^{-16}\,\mathrm{V\,cm^2}$. Vibrational distribution in the electronic ground state functions have been measured by different spectroscopic techniques in CO/He mixtures[37, 38]. A comparison between measured and calculated N_v distributions has been reported in Fig. 36 from Ref.[38].

v-v processes prevail over v-t ones and these distributions have practically no Boltzmann tail, a situation similar to that discussed for nitrogen in Sect. 3.1. Experimentally determined N_v distributions in CO/O_2 mixtures[17] have been reported in Fig. 37. The deactivating action of the oxygen atoms, formed in the dissociation of O_2, makes itself felt through the appearance of Boltzmann tail in N_v distribution.

Possible mechanisms of carbon monoxide dissociation in electrical discharges have recently been discussed in Ref.[39].

Rates of CO dissociation have been measured experimentally in Ref.[40]. At $\bar{u}_r = 0.5\,\mathrm{eV}$ and $n_e = 10^{10}\,\mathrm{cm^{-3}}$ the measured characteristic dissociation time $(k_d^e n_e)^{-1}$ is 30 ± 5 s.

Dissociation by DEM $e + CO\,(v = 0) \xrightarrow{k_d^e} e + C + O$ can definitely be ruled out for the experimental conditions reported. In fact calculation based on edf's of Ref.[34] and cross sections of Ref.[41] yields a value $(k_d^e n_e)^{-1} \sim 10^4$ s.

PVM can be very effective in pure CO under sufficiently strong e-v pumping (high n_e) because large v-v rates are associated with small v-t rates, like in N_2. A value of $k_d^s \cong 10\,\mathrm{s^{-1}}$ can be calculated for the following conditions: $E/N = 8 \times 10^{-16}\,\mathrm{V\,cm^2}$, $n_e = 10^{12}\,\mathrm{cm^{-3}}$, $N_{CO} = 10^{17}\,\mathrm{cm^{-3}}$ [39].

M. Capitelli and E. Molinari

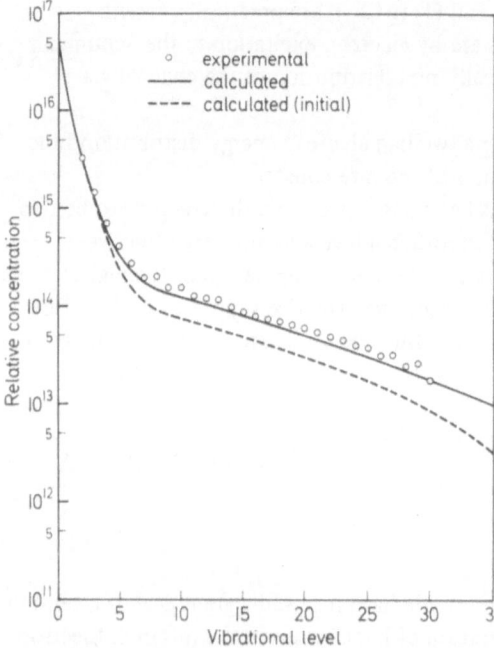

Fig. 36. Measured and calculated steady-state CO vibrational distributions for a CO/He mixture of 1/5 at 155 °K ($n_e = 5.8 \times 10^9$ cm^{-3}, $E/N = 2.7 \times 10^{-16}$ Vcm^2) (From Ref.[38])

Appropriate scaling to the experimental conditions of Ref.[40] would result in a characteristic time of 50 s. This value does however appear optimistic in view of a 3% dissociation producing O-atoms which depress k_d^s, as expected from the N_v distributions of Fig. 37. A more realistic model should make use of the JVE. An ap-

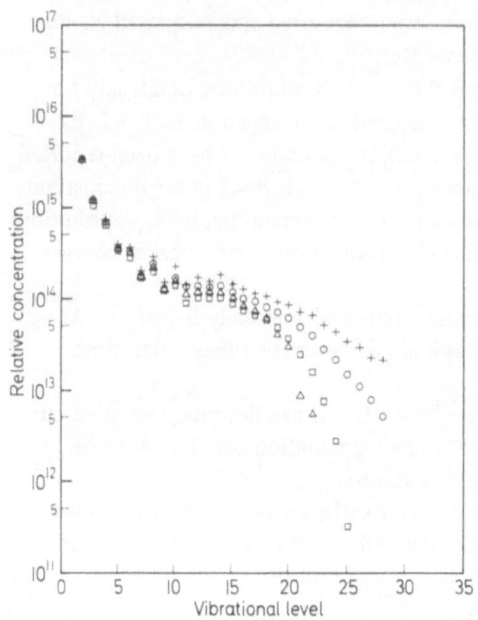

Fig. 37. Measured CO vibrational distributions for a 2./16.7 mbar CO/He mixture at 155 K and 10 mA for various O_2 additions. The symbols + ○ □ △ correspond to 0.0, 0.44, 0.66, 1 mbar of added O_2 (From Ref.[17])

proximate calculation based on a judicious estimate of $k_d^e(v)$ rates yield a dissociation rate two orders of magnitude larger than k_d^e ($v = 0$), i.e. a characteristic time of about 100 s, to be compared with an experimental one of 30 s. Further theoretical and experimental work is necessary on this system.

It should furthermore be pointed out that *chemical* dissociation channels such as

$$CO(v) + CO(v) \longrightarrow C_2O + O \quad \text{and} \quad CO(v) + CO(v) \longrightarrow CO_2 + C \tag{XL}$$

can be activated by the presence of vibrationally excited molecules[42]. Dissociation of carbon monoxide has also been obtained by pumping the vibrational system of CO with IR photons from a cw laser source[43, 44]. In this case N_v distributions and the possibility of CO dissociation can be explained on the basis of PVM; chemical channels remain open in this case as well.

3.4 Hydrogen Fluoride[45]

The efficiency of PVM in dissociating HF is linked to the possibility of creating a high vibrational temperature θ_1. This occurs when the pumping rate of the process

$$e + HF(v = 0) \xrightarrow{k_{01}^e} e + HF(v = 1) \tag{XLI}$$

overcomes the most important deactivation channel

$$HF(v = 1) + HF \xrightarrow{P_{10}^{HF}} HF(v = 0) + HF \tag{XLII}$$

Large θ_1 will thus be obtained when the inequality $n_e > \dfrac{P_{10}^{HF}}{k_{01}^e} N_{HF}$ holds. From this follows that appreciably high θ_1 can be expected at high values of n_e only. This is confirmed by Fig. 38a and 38b. Values of k_d^s are high for $n_e > 10^{12}$ cm^{-3} in agreement with the qualitative considerations presented above. The strong dependence of k_d^s on T_g is expected for reasons similar to those discussed for H_2.

Figure 39 shows typical N_v distributions at different times for $n_e = 10^{14}$ cm^{-3}. Also reported in this figure is the temporal evolution of the atoms. It should be noted that N_v distributions do not display a Boltzmann tail. In fact with $n_e = 10^{14}$ cm^{-3}, $T_g = 300$ K the vibrational temperature is high and v-v processes dominate over v-t processes. High electron concentrations of the order of 10^{14} cm^{-3} can be obtained in pulsed discharges with characteristic times of the order of 1 μs. Fig. 39 shows that the atom concentration reaches a value of 2×10^{15} cm^{-3} (i. e., about 1% dissociation) in a time of 10^{-6} s. HF dissociation induced by IR laser pumping has been investigated in Ref.[46, 47] with laser pulses of the order of 3×10^7 W/cm^2. The corresponding pumping rate for the process

$$h\nu + HF(v = 0) \longrightarrow HF(v = 1) \tag{XLIII}$$

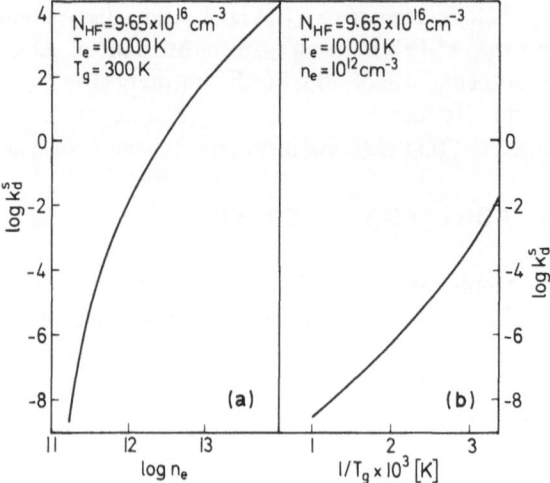

Fig. 38a and b. Values of k_d^s in HF as a function of (a) n_e (cm^{-3}) and (b) of $1/T_g$ (From Ref.[45])

is in this case about 10^9 s^{-1} and is able to overcome the vibrational deactivation of process (XLII) thereby introducing vibrational quanta in the HF system. v-v processes will transport these quanta up to the dissociation limit.

N_v distributions calculated in Ref.[46-47] for the laser pumped system are similar to those reported in Fig. 39. However, for the conditions, discussed above, $h\nu$-v pumping is more intense than e-v excitation and therefore laser pumping yields large k_d^s.

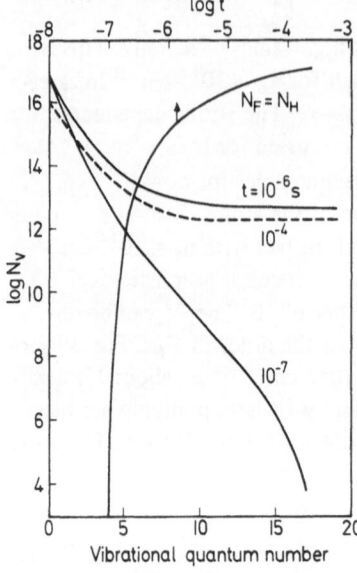

Fig. 39. Vibrational distribution N_v(cm^{-3}) in HF as a function of the vibrational quantum number at different times and atom concentration (cm^{-3}) as a function of time (s) [$n_e = 10^{14}$ cm^{-3}; $T_g = 300$ K; p = 4 mbar; T_e = 10000 K] (From Ref.[45])

4 Polyatomic Molecules

Three cases will be examined, which have formed the subject of recent experimental investigations, namely the dissociation of carbon dioxide[48], the cracking of hydro-carbons (CH_4, C_2H_6, C_2H_4, nC_4H_{10})[52], and ammonia decomposition[55]. Apparatus and conditions are, in these three cases, similar to those utilized for hydrogen and oxygen dissociation. Detailed informations on excitation and energy transfer processes, such as those available for diatomic molecules, are at present, insufficient to formulate a dependable model for the dissociation processes of these polyatomic molecules under plasma conditions. However, it will be shown that a simplified disso-ciation mechanism involving only direct excitation by electrons to excited dissocia-tive states can not account for the observed rates. A fraction of the energy pumped into the vibrational system of these molecules is apparently utilized for the dissocia-tion process and this fraction increases with increasing pressure.

4.1 The Dissociation of Carbon Dioxide[48]

Dissociative electron excitation of CO_2

$$CO_2 + e \longrightarrow CO_2^* + e \longrightarrow CO + O + e \qquad (XLIV)$$

involves the excited electronic state CO_2^*, approximately 7 eV above ground state.

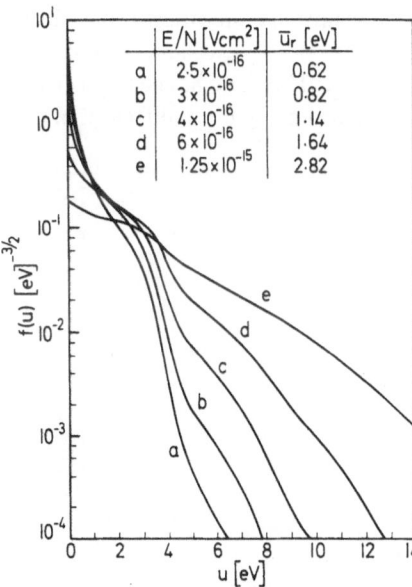

Fig. 40. Electron energy distribution functions in CO_2 for various E/N values (From Ref.[34])

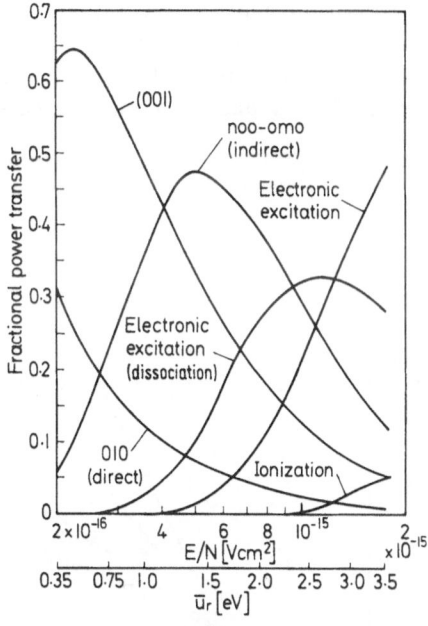

Fig. 41. Fractional electron power transfer in CO_2 as a function of E/N (From Ref.[34])

Cross sections for this process are available[49] and values of k_d^e for this process can then be calculated as a function of E/N by using the non-maxwellian edf's of Fig. 40 from Ref.[34]. Dissociative attachment according to

$$CO_2 + e \longrightarrow CO + O^- \qquad\qquad (XLV)$$

has been shown to make only a minor contribution to the dissociation process[48].

Figure 41 from Ref.[34] gives the fractional electron power transfer to the different excitation processes in pure CO_2. Fractional power transfer is a unique function of E/N and is also independent of both n_e and N_{CO_2}.

The initial rate of CO_2 dissociation according to reaction XLIV is[48]

$$- dN_{CO_2}/dt = k_d^e \, n_e \, N_{CO_2}^0 = F_d^e (E/N) \, \overline{W}/6.7 \times 10^5 \qquad\qquad (17)$$

$F_d^e (E/N)$ is the fractional power transferred to dissociative electronic excitation at a given E/N, as derived from Fig. 41, $\overline{W}$ is the electric power density transferred to the discharge (W cm^{-3}) and 6.7×10^5 J mol^{-1} is the 7 eV threshold energy of process XLIV, $N_{CO_2}^0$ is the initial number density of CO_2. The experimentally determined degree of CO_2 dissociation z_d at the exit of the discharge zone of length ℓ has been fitted by a first order equation with volume variation

$$\ell/\Phi = \frac{RT_g}{k_{exp}} \cdot \frac{1}{\pi r^2} \, A [-0.5 \, z_d - 1.5 \ln (1 - z_d)] \qquad\qquad (18)$$

with Φ, variable volumetric gas flow rate, r discharge tube radius, A conversion factor. Experimental plots of $- [0.5 \, z_d + 1.5 \ln (1 - z_d)]$ against ℓ/Φ are straight lines from which values of the experimental dissociation rate constant k_{exp} can be derived. The ratio $F_d^e(E/N) \, \overline{W}/6.7 \times 10^5 \, k_{exp} N_{CO_2}^0$ gives $k_d^e \, n_e/k_{exp}$. This ratio has been plotted in Fig. 42 as a function of the reduced average electron energy $\overline{u}_r$, at different pressures.

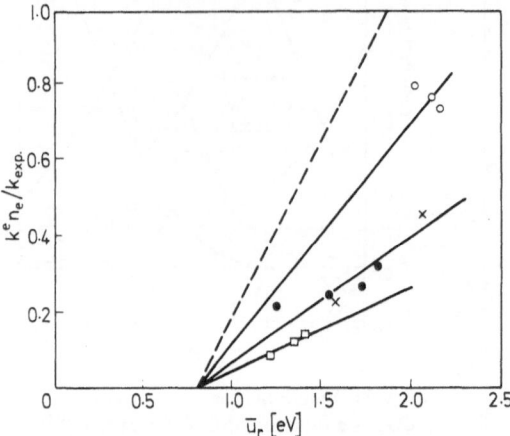

Fig. 42. $k_e n_e/k_{exp}$ as a function of $\overline{u}_r$ at different pressures in CO_2 (o ● □ 13.3, 26.7, 53.3 mbar; x and − − − see Ref.[48])

Table 3. Fraction of the energy pumped into the vibrational system of CO_2 utilized for the dissociation processes $(W_{vd})/(W_v)$, at different pressures and power densities (From Ref.[48])

p, mbar	$\bar{W}$, W cm^{-3}	$(W_{vd})/(W_v)$
53.3	15.7–23.	0.54–0.59
26.7	8.4–16.7	0.28–0.42
13.3	8.9–9.9	0.03–0.06

Measurements of initial rates of dissociation of CO_2 have also been performed in a lower pressure range (0.4–4 mbar)[50] (see also[51]). The dashed line of Fig. 42 gives the ratio between calculated $k^e_{dn_e}$ and the corresponding k_{exp} of Ref.[50]. The results reported in this figure show that process XLIV can not represent the unique mechanism of dissociation. The contribution of this process to the observed rates decreases with decreasing $\bar{u}_r$, becoming negligible around $\bar{u}_r = 0.8$ eV, and with increasing pressure. From Fig. 41 and Fig. 42 one finds that an increasing fraction of the power transferred to the vibrational modes of CO_2 should actually be utilized for dissociation when the pressure is increased. This fraction can easily be calculated and the results are given in Table 3 from Ref.[48]. One finds that at 53 mbar about 60% of the energy pumped into the vibrational systems of CO_2 is actually utilized in the dissociation process.

The contribution of chemical dissociative channels, activated by the presence of vibrationally excited CO_2, such as

$$CO_2(v) + CO_2(v) \longrightarrow CO + O + CO_2 \qquad \text{(XLVI)}$$
$$O + CO_2(v) \longrightarrow CO + O_2 \qquad \text{(XLVII)}$$

is discussed in Ref.[42]. Application of the theoretical treatment of Ref.[42] to the experimental data of Ref.[48] is not straightforward and needs closer attention.

4.2 Cracking of Hydrocarbons[52]

Cracking of CH_4, C_2H_6, C_2H_4, and n-C_4H_{10} has been investigated in the presence of H_2 (mostly equimolar mixtures of one hydrocarbon with hydrogen) at total pressures between 10 and 40 mbar. The results can be summarized as follows:
(1) The decomposition of all hydrocarbons investigated follows rather closely a *zero* order kinetics up to 80–90% conversion. (2) Rate constants of dissociation can be considered approximatively the *same for all hydrocarbons* studied (between 2 and 40 s^{-1}), depending on power input and hydrocarbon to hydrogen ratio and many orders of magnitude higher than the corresponding rates of "thermal" cracking (2×10^{-9} for CH_4, 2.6×10^{-3} s^{-1} for n-C_4H_{10}) at the maximum gas temperature

of the plasma (950 K). (3) The primary step for the decomposition is very likely to be the same for *all* hydrocarbons investigated and involves the rupture of a C–H bond. (4) The proposed reaction scheme is the following:

$$RH \xrightarrow{\text{k(XLVIII)}} R + H \qquad\qquad\qquad\qquad\qquad\qquad\qquad \text{(XLVIII)}$$

$$H_2 \xrightarrow{\text{k(XLIX)}} H + H \qquad\qquad\qquad\qquad\qquad\qquad\qquad \text{(XLIX)}$$

$$RH + H \xrightarrow{\text{k(L)}} R + H_2 \qquad\qquad\qquad\qquad\qquad\qquad\qquad \text{(L)}$$

$$R \longrightarrow \qquad C_2H_2 + polymer \qquad\qquad\qquad\qquad\qquad\qquad \text{(LI)}$$

Intermediate stable products

According to this simplified scheme the rate of disappearance of the hydrocarbons, when a stationary state for N_H is assumed, can be written as:

$$dN_{RH}/dt = 2k\,(XLVIII)N_{RH} + 2k\,(XLIX)N_{H_2} \qquad\qquad\qquad\qquad (19)$$

Since the concentration of RH decreases with time while that of H_2 increases as a result of hydrocarbon decomposition, a zero order kinetics can follow from Eq. (19) for appropriate values of the rate constants [$k(XLVIII) \cong k(XLIX)$]. Reactions grouped under (LI) in the kinetic scheme represent alternative routes for the conversion of radicals R ($CH_3, C_2H_5, C_2H_3, C_4H_9$) into C_2H_2, which is the main reaction product, and into polymer. Intermediate stable products are C_2H_4, CH_4 and C_3H_6. (5) Rate constants $k(XLVIII)$ are significantly larger than those predicted

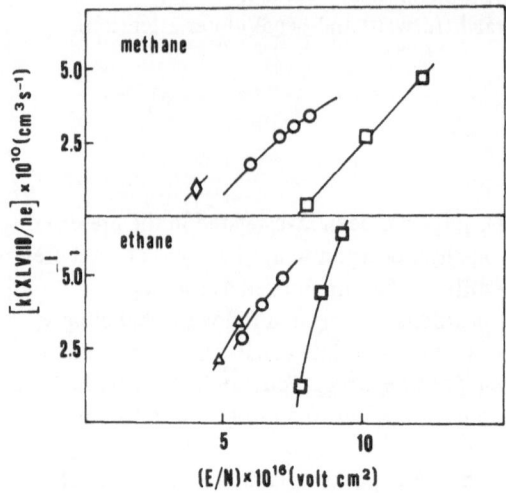

Fig. 43. Values of $k(XLVIII)/n_e$ for CH_4 and C_2H_6 as a function of E/N at 13.3, 26.7 and 53.3 mbar (From Ref.[52])

by a mechanism of direct electron impact dissociation RH + e $\longrightarrow$ R + H + e and imply the presence of additional mechanisms involving vibrotationally excited molecules. This is well illustrated by the data of Fig. 43[52] where rate constants k(XLVIII) for CH_4 and C_2H_6 decomposition, divided by the estimated electron concentrations n_e, have been plotted against E/N at different pressures. Dissociation by direct electron impact implies a rate constant $k_d^e = k(XLVIII)n_e$ which should be a unique function of E/N. The results given in the figure show that this mechanism can not be the only one operating under the reported experimental conditions. In fact, even on the assumption that at the lowest pressure investigated dissociation by direct electron impact is predominant, the contribution of this mechanism decreases with increasing pressure and becomes negligible at 26.7 mbar. These results closely parallel those previously described for hydrogen and carbon dioxide.

Sources of previous work on hydrocarbon cracking can be found in Ref.[52]. Cracking of CH_4 and of C_4H_{10} has also been studied in the presence of CO_2 and H_2O vapor[53, 54]. The interesting observation is that the rupture of a C–H bond remains the slow step which occurs at rates very close to those measured in mixtures with hydrogen. Further oxidation to CO follows via C_2 species. In the presence of O_2[52], conversion to CO occurs through self-accelerated radical chain reactions.

4.3 The Decomposition of Ammonia[55]

NH_3 decomposition has been studied in the pressure range 7–53 mbar with simultaneous spectroscopic observation of emission bands of N_2, NH, H_2 and of atomic lines of H.

Spectroscopic observations confirm the existence of a vibrational temperature $T_v \cong 2000$ K greater than the translational one ($T_g \cong 800$ K) and show that NH is an important intermediate. The decomposition follows a zero order kinetics in analogy with the results reported in Sect. 4.2. Values of the rate constants are again several orders of magnitude larger than those observed in thermal decomposition and do not differ appreciably from those derived for hydrocarbons under similar discharge conditions. The reaction scheme is similar to that proposed in Sect. 4.2. The slow step involves the rupture of a N–H bond and the final step is NH + NH$\longrightarrow$ N_2 + H_2. The presence of a mechanism of dissociation which makes use of the energy pumped into the vibrational manifold of NH_3 has been invoked for this system as well.

5 Comments

In previous sections an attempt has been made to compare available experimental rates of selected plasmochemical reactions with predictions by mechanisms (PVM, DEM, JVE) based on the existence of electronic and vibrational nonequilibrium in the plasma.

While N_v distributions find increasing support from direct and indirect experimental observations, the comparison of predicted and observed reaction rates suffers of limitations which are both experimental and theoretical.

On the experimental side, limited accuracy concerns the evaluation of n_e, T_g, E/N, $f(u)$ as well as of the fluid dynamics of the chemical discharge reactor i.e., type of flow (plug, laminar, turbulent) and related dimensionless quantities relevant to the definition of (a) velocity profiles and corresponding effective residence times, (b) of transport properties, needed in the calculation of deexcitation of excited species and of recombination of atoms on the walls, of heat transfer to the walls, and of back-mixing[58]. For example the constriction of the plasma column is a complicating factor at pressure above about 5 mbar.

The situation on the theoretical side requires a more detailed analysis:

The model mechanism, discussed in the preceeding sections, requires the following refinements: (a) A complete temporal coupling between the system of master equations and the relevant Boltzmann equation should always be made in order to obtain edf's and N_v distributions consistent with each other and with the presence of atomic species. The recombination of the atoms should be taken into account. All this can be done but represents a serious mathematical and computational problem. (b) Dissociation mechanisms are particularly sensitive to different excitation and transfer processes, as illustrated in the preceeding sections. More accurate cross sections are needed. In fact differences of one or more orders of magnitude are not unusual among literature values. New excitation channels might come into play which modify calculated rates. For example recent work[56] indicates the existence of resonances in H_2^- which could substantially increase e-v cross sections of high vibrational levels. JVE in H_2 should also take into account the possibility of dissociation over triplet states higher than $b^3 \Sigma_u^+$ state normally considered in the calculations. Recent work[57] has in fact shown that dissociation channels of comparable importance are those involving electronic excitation to $a^3 \Sigma_g^+$ and $c^3 \Pi_u$. This point has partially been considered at the end of Sect. 2.3.4.

The mechanisms reported in the previous sections refer to a simplified model of the plasma. This point will be illustrated in the sections to follow.

5.1 The Role of Walls

The possibility of the deactivation at the reactor walls of vibrationally excited molecules was not considered. This contribution to the overall loss of excitation is determined by the accomodation coefficient for vibrational energy of the wall material and by the transport coefficients of the molecules to the walls. An example will serve for illustration: The first order rate constant k_w which accounts for the losses of vibrational excitation of an excited species at the reactor walls can be expressed as

$$k_w = \frac{2}{r} \, k_w' \delta / (k_w' + \delta) \tag{20}$$

Table 4. Values of k_w (s^{-1}) for H_2 and N_2

H_2			N_2	
α_d	a	b	a	b
10^{-3}	118.5	92.2	38.0	31.1
10^{-4}	26	15.3	7.2	4.2

a: $p = 26.7$ mbar, $\Phi = 7.4 \times 10^{-3}$ mols^{-1}, $r = 2.4$ cm, $T_g = 1000$ K
b: $p = 1.33$ mbar, $\Phi = 7.4 \times 10^{-4}$ mols^{-1}, $r = 2.4$ cm, $T_g = 300$ K

Values of α_d $10^{-3}–10^{-4}$ have been reported for $N_2(v)$ on glass by Polak
(Pure Appl. Chem. 39, 307, 1974).

where $k'_w = \frac{1}{4} c \alpha_d$, with c random velocity of the species at T_g and α_d deexcitation
coefficient for vibrational energy at the wall. δ is given by $\delta = Sh \frac{D}{2r}$ where r is discharge tube radius, Sh, Sherwood number for cylindrical geometry, which is a func-
tion of the Schmidt and Reynold's numbers[58] and D is the diffusion coefficient,
k_w represents the rate constant for a 1st order heterogenous wall process in the presence of limitations by diffusion[59]. Values of the Sherwood number depend on type
of flow (plug, laminar) and gas flow rate and this introduces a dependence of k_w
on flow conditions. Table 4 gives some of the calculated k_w values for conditions typical of the experiments carried out in our laboratory. One appreciates that vibrational
deactivation at the walls can be very important for H_2, since k_w is of the same order
of magnitude of the e-v pumping rate $(n_e k^e_{01} \sim 200$ s^{-1}, $n_e = 10^{11}$ cm^{-3},
$E/N > 2 \times 10^{-16}$ V cm^2). A smaller influence is expected for N_2 as a consequence
of higher e-v pumping rates (of the order of 10^3 s^{-1}) and of lower k_w values. The
influence of the deexcitation coefficient on calculated k_w should also be noted.

Wall deactivation of vibrationally excited nitrogen should however be considered
at number density lower than 10^{15} cm^{-3} as pointed out in Ref.[7]. The influence on
the kinetics of hydrogen dissociation of the transport of atoms to the walls and their
recombination thereon has been considered in Ref.[16]. Calculated values of k_w are
affected by the scattering in the literature values of the recombination coefficients.

5.2 The Role of Metastables

Figure 44[60] gives a simplified energy level scheme for N_2 which includes the ground
state singlet $N_2(X^1\Sigma^+_g)$ and the triplets $N_2(A^3\Sigma^+_u)$ $N_2(B^3\Pi_g)$ and $N_2(C^3\Pi_u)$. Electronically excited metastable states and the ground state are joined in this figure by
heavy lines indicating the main connecting processes, process of secondary importance being shown as broken lines. Rate constants for these processes have been evalu-

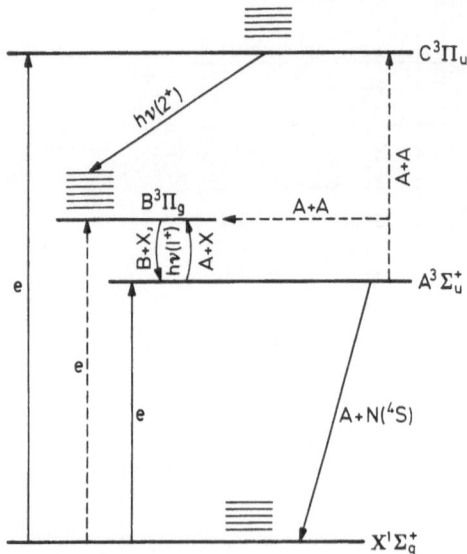

Fig. 44. Simplified energy diagram of the nitrogen system (From Ref.[61])

ated in[60, 61] and in references given therein. Beside excitation by direct electron impact from the ground state (vertical arrows) important processes for the population of the various states are the following

$$2N_2(A^3\Sigma_u^+) + N_2(X^1\Sigma_g^+) \longrightarrow N_2(C^3\Pi_u) + 2N_2(X^1\Sigma_g^+) \tag{LII}$$

$$N_2(A^3\Sigma_u^+) + N_2(X^1\Sigma_g^+\ v > 20) \longrightarrow N_2(C^3\Pi_u) + N_2(X^1\Sigma_g^+\ v\text{-}\Delta v) \tag{LIII}$$

which populate the C state

$$N_2(A^3\Sigma_u^+) + N_2(X^1\Sigma_g^+\ v \geq 6) \longrightarrow N_2(B^3\Pi_g) + N_2(X^1\Sigma_g^+\ v\text{-}\Delta v) \tag{LIV}$$

$$2N_2(A^3\Sigma_u^+) \longrightarrow N_2(B^3\Pi_g) + N_2(X^1\Sigma_g^+) \tag{LV}$$

which populate the B state.

$N_2(A^3\Sigma_u^+)$ is populated from the upper B level. An important deactivation process is $N_2(A^3\Sigma_u^+) + N(^4S) \longrightarrow N_2(X^1\Sigma_g^+) + N(^2P)$. Steady state concentrations of the various electronic states will therefore be dependent on the edf and on the N_v distribution as well as on pressure and gas temperature. Dissociation of N_2 can take place from any one of these excited electronic states by PVM, DEM or JVE, beside dissociation from the ground $N_2(X^1\Sigma_g^+)$ state which is the only one considered in our simplified model. However, the extent of the contribution of triplet states to the overall rate of dissociation can not be evaluated with presently available data.

Results of a detailed spectroscopic investigation of a glow discharge in nitrogen at low pressure have recently been reported in Ref.[62]. In this work concentration of $C^3\Pi_u$, $B^3\Pi_g$ and $A^3\Sigma_u^+$ species have been determined at different radial positions in the discharge, together with the corresponding distributions of vibrational populations. These results have satisfactorily been intepreted along the lines outlined above by using appropriate edf's. Population distributions among vibrationally levels in the ground $X^1\Sigma_g^+$ state have also been derived. The connection between vibrational temperatures in the $C^3\Pi_u$ and in the ground $X^1\Sigma_g^+$ state has been established in discharges operated in the moderate and high pressure range in Ref.[63].

The dissociation of oxygen, discussed in Sect. 3.2.2, is an instructive example of a predicted contribution of the $(a^1\Delta_g)$ and $(b^1\Sigma_g^+)$ metastables to the dissociation process which has actually been invoked in low pressure discharges[30, 31] and which is absent under the discharge conditions discussed in Sect. 3.2.2. as a consequence of mechanisms competing with the excitation of these metastables to the dissociative electronic states. Under the experimental conditions of Ref.[32] a decisive role is apparently played by the high lying metastables $(C^3\Delta_u, A^3\Sigma_u^+, C^1\Sigma_u^-)$ which increase the net rate of dissociation by a substantial reduction of the recombination rate of O-atoms.

5.3 The Role of Rotational Excitation

In the discussion of the electronic and vibrational disequilibrium no mention has been made of the possible role of rotational excitation. This was justified by the fact that, for values of $\bar{u}_r > 0.5$ eV, the contribution of rotational excitation to the edf can be neglected and that the fractional energy transferred by the electrons to rotation is also negligible[34].

However, the data of Fig. 23 and the values of T_r measured at 5–50 mbar in various systems (gas NH_3/H_2, emitting species NH, N_2, T_r = 2000 K[55]; gas CO, emmitting species C_2, T_r = 6000–8000 K[64]) indicate that, in contrast to the situation prevailing at pressures below 5 mbar, where $T_r \longrightarrow T_g$, above 5 mbar T_r increases with increasing pressure and becomes equal to T_v at about p = 15 mbar. Rotational distributions are apparently Boltzmann in these discharges[19, 20] and rotational levels of very high rotational quantum number are appreciably populated at the measured T_r's.

In a recent paper[65], devoted to energy transfer processes in HF it has been shown by trajectory calculations, that v-r processes of the type

$$HF(v_1 = 1, J_1 = 2) + HF(v_2 = 0, J_2) \longrightarrow HF(v_1 = 0, J_1 = 10-16) + HF(v_2 = 0, J_2)$$

$$(LVI)$$

$$HF(v_1 = 3, J_1 = 2) + HF(v_2 = 0, J_2) \longrightarrow HF(v_1 = 2, J_1 = 10-16)$$
$$\longrightarrow HF(v_1 = 1, J_1 = 16-20) \left. \right\} + HF(v_2 = 0, J_2)$$
$$\longrightarrow HF(v_1 = 0, J_1 = 21-26)$$
$$(LVII)$$

proceed at very fast rates, with a probability of $\simeq 0.2$ and $\simeq 0.1$ for single-quantum and multiple-quantum transitions, respectively. In these processes the vibrational energy of the vibrationally excited incident HF is transferred by an intramolecular process into rotational energy of the same molecule, with practically no change of the internal energy state of the target HF molecule. One appreciates form Eq. (LVII) that HF (v = 0) species are formed, and that therefore vibrational energy can be relaxed from high v states by these additional fast processes. Very high rotational states are populated via v-r processes with much smaller energy defects than would have been predicted if both reagent and product HF species were assumed rotationless.

The reverse processes of (LVI) and (LVII), which proceed at similar rates, represent additional channels for the pumping up of vibrational energy.

The occurrence of v-r processes of this kind can, in principle, have important consequences for both the PVM and the JVE mechanisms. This can be illustrated by the case of hydrogen. A number of resonant or near-resonant levels can be found in both para and ortho hydrogen, e.g.,

$$H_2 - p \, (v = 0, J = 8) \text{ and } H_2 - p \, (v = 1, J = 0)$$

$$H_2 - o \, (v = 0, J = 9) \text{ and } H_2 - o \, (v = 1, J = 3)$$

and others at $J \geqslant 8, 9$. The mole fraction of para and ortho molecules with J values $\geqslant 8, 9$ is a function of the rotational temperature, as given below:

$T_r(K)$	4000	2000	1000	500
mole fraction	0.2	0.05	0.003	10^{-5}

The rate of up-pumping ($v = 0 \longrightarrow v = 1$) can be written as $\bar{k}_{r\text{-}v} \, N_{J \geqslant 8,9} N_{H_2}$, where $\bar{k}_{r\text{-}v}$ is the weighted average of the r-v rate constants for each $v = 0, J \geqslant 8,9$ level. With overall transfer probabilities of the order of 0.3[65] this rate can be larger by one to three orders of magnitude at $1000 \, K < T_r < 4000 \, K$ than e-v (0,1) rates utilized in the previous sections for establishing the vibrational temperature θ_1. These considerations can be extended to $v \longrightarrow v + 1$ ($v > 0$) pumping. A theoretical evaluation of the N_v distributions which determine the rate constants k_d^s and k_{dj}^s of the PVM and of the JVE mechanism respectively, should therefore include direct and reverse v-r processes in the master equations. This is a very complex problem which remains to be solved. Qualitatively speaking the rate of up-pumping depends on T_r, mostly through $N_{J > J \, min}$ and on N_0^2. When T_r depends on pressure as indicated by Figure 25 one expect a rapid increase of this rate between about 5 and 15 mbar, followed by a slower increase with pressure. This behaviour is reflected, in different ways, by k_d^s and k_{dj}^s and one therefore expects that the ratio $k_d^e n_e / k_{exp}$ should decrease with increasing pressure, in accordance with the experimental evidence for H_2 (Fig. 18), CO_2 (Fig. 42), CH_4 and C_2H_6 (Fig. 43) and probably for NH_3. On the contrary, rotationless PVM and JVE mechanism decrease their contribution to the total rate of dissociation with increasing pressure and gas temperature (Sect. 2).

There is another important consequence of the inclusion of rotational excitation. In a rotationless model, the dissociation pseudo-level is taken at $v' + 1$ ($v' = 14$ for H_2). When high J levels are excited this pseudo level can be associated with a lower v[66], and this results in a strong increase of k_d^s (Sect. 2.3.1).

5.4 Concluding Remarks

In spite of the uncertainities involved in a comparison of calculated and observed dissociation rates which have been illustrated in the preceeding sections, the following conclusions appear to be justified by the analysis which has been presented in this chapter.

1) Hydrogen. JVE is the dominant mechanism at pressures below about 5 mbar; the contribution of PVM becomes appreciable below T_g = 300 K. In the range of moderate pressures (5–55 mbar) these mechanisms appear to be insufficient to account for the observed decrease of the ratio $k_d^e n_e / k_{exp}$ with increasing pressure. The contribution of rotational excitation to k_{dj}^s is a possible suggestion which could account at least qualitatively for the pressure dependence of the $k_d^e n_e$ ratio.

2) Nitrogen. PVM rates can be very high for pure nitrogen. The presence of different gases (hydrogen-oxygen) and/or N-atoms, as well as the effect of the walls, reduce the importance of this mechanism which should then be replaced by JVE. The role of metastables has not been ascertained quantitatively but can, in principle, be important.

3) Oxygen. As a consequence of the low *e-v* pumping rates and of vibrational deexcitation by O-atoms, this system appears to be dominated by DEM both at low and intermediate pressures. The predicted strong contribution of metastable $O_2(^1\Delta)$ is suppressed in the range of moderate pressures. Metastables close to the dissociation limit are likely to be responsible for the reduction of the rate of O-atom recombination.

4) Carbon monoxide. Considerations similar to those presented for nitrogen are valid for this system as well.

5) Hydrogen fluoride. Dissociation of this molecule requires very high n_e (pulsed discharges).

6) Polyatomic molecules (CO_2, hydrocarbons, NH_3). DEM contribution decreases in all cases with increasing pressure and decreasing E/N. An increasing fraction of the energy pumped into the vibrational systems of these molecules is utilized for dissociation when the pressure is increased. There are no available PVM or JVE calculations for these systems.

The theoretical analysis of non-equilibrium plasmas, developed in this chapter, appears therefore to provide a useful framework into which experimental observations on plasmochemical dissociation can be fitted. This is generally true at pressures below about 5 mbar, where T_r is not very different from T_g. At higher pressures a more sophisticated analysis must be developed which includes excitation and non-equilibrium of rotational levels and which might hopefully help understanding the behaviour reported at moderate pressures. Chemical dissociation channels activated by the presence of vibrotionally excited species should also be duly considered. It should finally be remembered that the pure vibrational mechanism discussed for plasmochemical dissociation can be extended to dissociation induced by infrared single-photon laser pumping after substitution of *e-v* excitation by *hv-v* pumping (Sect. 3.3 and 3.4). Infrared single photon laser pumping is a topic of current research activity[67] which is becoming progressively entangled with plasmochemical processes.

M. Capitelli and E. Molinari

Acknowledgements. The authors wish to express their indebtedness to Prof. F. Cramarossa and to the staff of the Centro di Studio per la Chimica dei Plasmi del C.N.R. (Drs. M. Cacciatore, P. Capezzuto, R. d'Agostino, M. Dilonardo, G. Ferraro, C. Gorse) for their contribution to a lively discussion of the subjects presented in this chapter.

6 References

1. Principles of Laser Plasmas, Bekefi, G. (ed.) New York: Wiley 1976
2. Capitelli, M., Dilonardo, M.: Z. Naturforsch. *34a*, 585 (1979) see also 4th Europhysics Study Conference on Atomic and Molecular Physics of Ionized Gases (ESCAMPIG), Essen 1978, paper C11
3. Michel, P., Pfau, S., Rutscher, A., Winkler, R.: Proceedings of XIII Conference on Phenomena in Ionized Gases, Berlin 1977, contributed papers p. 247.
4. Capitelli, M., Dilonardo, M., Molinari, E.: Chem. Phys. *20*, 417 (1977)
5. Capitelli, M., Dilonardo, M.: Chem. Phys. *30*, 95 (1978)
6a. Capitelli, M., Dilonardo, M.: Chem. Phys. *24*, 417 (1977)
6b. Capitelli, M., Dilonardo, M.: Rev. Phys. Appl. *13*, 115 (1978)
7. Polak, L. S.: Pure and Appl. Chem. *39*, 307 (1974)
8. Gordiets, B. F., Mamedov, S. S., Shelepin, L.: Sov. Phys. JETP *40*, 640 (1975)
9. Cacciatore, M., Capitelli, M., Dilonardo, M.: Chem. Phys. *34*, 193 (1978)
10. Treanor, C. E., Rich, J. W., Rehm, R. G.: J. Chem. Phys. *42*, 1798 (1968)
11. Kewley, D. J.: J. Phys. B.: At. Mol. Phys. *8*, 2565 (1975)
12. Capitelli, M., Ficocelli, E., and Molinari, E.: Equilibrium compositions and thermodynamic properties of Ar-H_2 plasmas. C.S.C.P. C.N.R. Bari (1972); see also Capitelli, M. and Ficocelli, E.: Rev. Int. Hautes Temp. et Refr. *14*, 195, 1977
13. Heidner, R. F., Kasper, J. V. V.: Chem. Phys. Lett. *15*, 179 (1972)
14. Shirley, J. A., Hall, R. J.: J. Chem. Phys. *67*, 2419 (1977)
15. Bell, A. T.: Ind. Eng. Chem., Fundam. *11*, 209 (1972)
16. Capezzuto, P., Cramarossa, F., d'Agostino, R., Molinari, E.: J. Phys. Chem. *79*, 1487 (1975)
17. Fisher, E. R., Lightman, A. J.: J. Appl. Phys. *49*, 530 (1978)
18. Anketell, J., Brockleurst, B.: J. Phys. B: Atom. Molec. Phys. *7*, 1937 (1974)
19. Cramarossa, F., Ferraro, G. Molinari, E.: J. Quant. Spectrosc. Radiat. Transfer. *18*, 471 (1977)
20. Cramarossa, F., Ferraro, G. Molinari, E.: J. Quant. Spectrosc. Radiat. Transfer. *14*, 419 (1974). Cramarossa, F., Ferraro, G.: J. Quant. Spectrosc. Radiat. Transfer. *14*, 159 (1974)
21. Shaub, W. M., Nibler, J. W., Harvey, A. B.: J. Chem. Phys. *67*, 1883 (1977)
22a. Polak, L. S., Sergeev, P. A., Slovetsky, D. I., Todesaite, R. D.: in Proceedings 12th Int. Conf. Phenomena in Ionized Gases, Part. 1; Eindhoven, 1975. Hölscher, J. G. A., Schram, D. C. (eds.) Amsterdam: North-Holland P. C. 1975
22b. Polak, L. S., Sergeev, P. A., Slovetsky, D. I., 4th ESCAMPIG, Essen '78 paper C26
23. Cacciatore, M., Capitelli, M., Dilonardo, M.: Beitr. Plasma Phys. *18*, 279 (1978)
24. Kiefer, J. H.: J. Chem. Phys. *57*, 1938 (1972)
25. Webster III, H., Bair, E. J.: J. Chem. Phys. *56*, 6104 (1973)
26. Francis, P. D.: Brit. J. Appl. Phys. *2*, 1717 (1969)
27. Kaufmann, F.: In: Chemical Reactions in Electrical Discharges, Gould, R. F. (ed), Adv. in Chem. Series *80*, (1969)
28. Mearns, A. M., Morris, A. J.: Chem. Eng. Progr. Symp. Ser. *67*, 37 (1971)
29. Bell, A. T., Kwong, K.: A.I.Ch.E. *18*, 990 (1972); Ind. Eng. Chem. Fundam. *12*, 90 (1973)
30. Sabadil, H., Biborosh, L. Koebe, G.: Beitr. Plasma Phys. *15*, 319 (1975). Sabadil, L., Zielke, E.: In: Proceedings 13th Int. Conf. Phenomena in Ionized Gases; Part 1. Berlin 1977
31. Yaron, M., Von Engel, A.: Chem. Phys. Letters *33*, 316 (1975); 3rd Int. Symp. Plasma Chemistry, Paper G.2.24, Limoges 1977
32. Cramarossa, F., d'Agostino, R., Molinari, E.: Beitr. Plasma Phys. *18*, 301 (1978)
33. Mašek, K., Růžička, T., Láska, L.: Czech. J. Phys. B. *27*, 888 (1971)
34. Nighan, W. L.: Phys. Rev. *A 5*, 1989 (1970)

35. Rockwood, S. D., Brau, J. E., Proctor, W. A., Canavan, G. H.: I.E.E.E.J. Quantum Electronics *QE9,* 120 (1973)
36. Lowell Morgan, W., Fisher, E. R.: Phys. Rev. *A 16,* 1186 (1977)
37. Brechignac, P., Martin, J. P., Taieb, G.: I.E.E.EJ. Quantum Electronics *Q E10,* 797 (1974)
38. Lightman, A. J., Fischer, E. R.: J. Appl. Phys. *49,* 971 (1978)
39. Schmailzl, U., Capitelli, M.: Chem. Phys. *41,* 143 (1979)
40. D'Amico, K. M., Smith, A. C. S.: J. Phys. D: Appl. Phys. *10,* 261 (1977)
41. Novgorodov, M. Z., Sobolev, N. N.: Proceedings 11th Int. Conf. Phenomena in Ionized Gaeses, Invited papers, Prague 1973
42. Legasov, V. A., Rusanov, V. D., Fridman, A. A., Sholin, G. V.: 3rd Int. Symp. on Plasma Chemistry, paper 6.5.18, Limoges 1977
43. Rich, J. W., Bergman, R. C.: Calspan Report No. WG-6005.A.1 April 1977, Calspan Corporation, Buffalo New York 14221
44. Rich, J. W., Bergmann, R. C., Raymonda, J. W.: Appl. Phys. Lett. *27,* 656 (1975)
45. Capitelli, M., Dilonardo, M.: Z. Naturforsch. *33a,* 1085 (1978)
46. Pummer, H., Proch, D., Schmailzl, U., Kompa, K. L.: J. Phys. D: Appl. Phys. *11,* 101 (1978)
47. Schmailzl, U., Pummer, H., Proch, D., Kompa, K. L.: J. Phys. D: Appl. Phys. *11,* 111 (1978)
48. Capezzuto, P., Cramarossa, F., d'Agostino, R., Molinari, E.: J. Phys. Chem. *80,* 882 (1976)
49. Hake, R. D., Phelps, A. V.: Phys. Rev. *158,* 70 (1967)
50. Kutsegi Corvin, K., Corrigan, S. J. B.: J. Chem. Phys. *50,* 2570 (1969)
51. Brown, L. C., Bell, A. T.: Ind. Eng. Chem. Fundam. *13,* 203, 210 (1974)
52. Capezzuto, P., Cramarossa, F., d'Agostino, R., Molinari, E.: Beitr. Plasma Phys. *17,* 205 (1977)
53. Capezzuto, P., Cramarossa, F., d'Agostino, R., Molinari, E.: Rev. Phys. Appl. *12,* 1205 (1977)
54. Capezzuto, P., Cramarossa, F., d'Agostino, R., Molinari, E.: Combustion and Flame *33,* 251 (1978)
55. Cramarossa, F., Colaprico, V., d'Agostino, R., Molinari, E.: 3rd Int. Symp. on Plasma Chemistry, Paper G.5.13 Limoges 1977
56. Gresteau, F., Hall, R. I., Mazeau, J., Vichon, D.: J. Phys. B: At. Mol. Phys. *10,* L 545 (1977)
57. Chung, S., Lin, C. C.: Phys. Rev. *A 17,* 1874 (1978)
58. Franck-Kamenetskii, D. A.; Diffusion and heat transfer in chemical kinetics, New York: Plenum Press 1969
59. Bennet, C. O., Meyers, J. E.: Momentum, heat and mass transfer. New York: Mc Graw Hill, 1962
60. Polak, L. S., Slovetsky, D. I., Sokolov, A. S.: 3rd. Int. Symp. on Plasma Chemistry, Paper G.5.4., Limoges 1977
61. Polak, L. S., Slovetsky, D. I., Urbas, A. D.: 3rd Int. Symp. on Plasma Chemistry, Paper G.5.3, Limoges 1977
62. Behnke, J. F., Grigull, P., Scheiber, H., Report 1977/78, Sektion Physik Elektronik, Universität, Greifswald, (DDR), 1978
63. Locqueneux Lefebvre, M., Ricard, A.: Rev. Phys. Appl. *12,* 1213 (1977)
64. Unpublished work from this laboratory
65. Wilkins, R. L.: J. Chem. Phys. *67,* 5838 (1977); Kwok, M. A., Gross, E. F., Wilkins, R. L.: 10th Int. Quantum Electronics Conf., Paper H.11, Atlanta 1978)
66. Valance, W. G., Schlag, E. W., Elwood, J. P.: J. Chem. Phys. *47,* 3284 (1967)
67. Chemical and biochemical applications of lasers, Bradley Moore, C. (ed.), New York: Academic Press 1974, 1977
68. Engelhardt, A. G., Phelps, A. V.: Phys. Rev. *131,* 2115 (1963)
69. Rockwood, S. D., Phys. Rev. *A 8,* 2348 (1973)
70. A very rich bibliography on edf can be found in a) Rutscher, A., Proceedings of XIII Int. Conf. on Phenomena in Ionized Gases, Berlin 1977 (Invited papers); b) Whilhelm, J., Winkler, R., Proceedings of XIV Int. Conference on Phenomena in Ionized Gases, Grenoble 1979 (Invited papers)
71. Hertzfeld, H. F., Litowitz, T. A.: Absorption and dispersion of ultrasonic waves. New York: Academic Press 1959

Received May 21, 1979

Subject Index

111

Subject Index

Fossil fuels   2ff

Gilsonite   46

HCN   6, 13ff, 28, 30ff, 40, 41, 46, 48
HF   61, 95ff, 105
H_2 5ff, 15, 20ff, 35ff, 61, 64ff, 70, 100, 106
H_2^+ 7, 8
H_3^+ 8
H_2/N_2 80
H_2O 26ff, 41, 44, 46, 48
H_2S 22, 45

Ionization   7, 65

Laser   60, 93, 107
– decomposition of coal 39
– – of crude oil   22
– – of natural gas   14

Master equation   67ff, 74ff, 106
Metastables   103ff
Methane plasma   4
– ionic composition   7

N_2 13ff, 18, 26, 35, 38, 41, 61, 78ff, 103ff
Natural gas plasma   4
NH_3 101
NH_3/H_2 105
Nitrogen arc in methane   14

O_2 26, 61, 82ff
O_3 90

Petroleum   14
– plasma desulfurization of   22
Plasma jet   10, 17, 34
Plasma pyrolysis
– coal of   27, 36
– crude oil of   18
– gasoline of   17
– methane of   13
– tar, oil shales   46
Plug flow plot   91
Polyatomic molecules   97ff
Population density – see "Energy distribution"
Power transfer   66, 97, 98

Recombination   73, 74, 83
– assisted dissociation   84, 92

Temperature   60ff
– electrons of – see "el. energy distribution"
– gas of   77
– rotational   81, 106
– vibrational   81
Treanor distribution   70, 87

Walls of reactor   102ff

Author Index Volumes 26–90

The volume numbers are printed in italics

Adams, N. G., see Smith, D.: *89*, 1–43 (1980).

Albini, A., and Kisch, H.: Complexation and Activation of Diazenes and Diazo Compounds by Transition Metals. *65*, 105–145 (1976).

Altona, C., and Faber, D. H.: Empirical Force Field Calculations. A Tool in Structural Organic Chemistry. *45*, 1–38 (1974).

Anderson, D. R., see Koch, T. H.: *75*, 65–95 (1978).

Anderson, J. E.: Chair-Chair Interconversion of Six-Membered Rings. *45*, 139–167 (1974).

Anet, F. A. L.: Dynamics of Eight-Membered Rings in Cyclooctane Class. *45*, 169–220 (1974).

Anh, N. T.: Regio- and Stereo-Selectivities in Some Nucleophilic Reactions. *88*, 145–162 (1980).

Ariëns, E. J., and Simonis, A.-M.: Design of Bioactive Compounds. *52*, 1–61 (1974).

Ashfold, M. N. R., Macpherson, M. T., and Simons, J. P.: Photochemistry and Spectroscopy of Simple Polyatomic Molecules in the Vacuum Ultraviolet. *86*, 1–90 (1979).

Aurich, H. G., and Weiss, W.: Formation and Reactions of Aminyloxides. *59*, 65–111 (1975).

Balzani, V., Bolletta, F., Gandolfi, M. T., and Maestri, M.: Bimolecular Electron Transfer Reactions of the Excited States of Transition Metal Complexes. *75*, 1–64 (1978).

Bardos, T. J.: Antimetabolites: Molecular Design and Mode of Action. *52*, 63–98 (1974).

Barnes, D. S., see Pettit, L. D.: *28*, 85–139 (1972).

Bauder, A., see Frei, H.: *81*, 1–98 (1979).

Bastiansen, O., Kveseth, K., and Møllendal, H.: Structure of Molecules with Large Amplitude Motion as Determined from Electron-Diffraction Studies in the Gas Phase. *81*, 99–172 (1979).

Bauer, S. H., and Yokozeki, A.: The Geometric and Dynamic Structures of Fluorocarbons and Related Compounds. *53*, 71–119 (1974).

Baumgärtner, F., and Wiles, D. R.: Radiochemical Transformations and Rearrangements in Organometallic Compounds. *32*, 63–108 (1972).

Bayer, G., see Wiedemann, H. G.: *77*, 67–140 (1978).

Bernardi, F., see Epiotis, N. D.: *70*, 1–242 (1977).

Bernauer, K.: Diastereoisomerism and Diastereoselectivity in Metal Complexes. *65*, 1–35 (1976).

Bikerman, J. J.: Surface Energy of Solids. *77*, 1–66 (1978).

Birkofer, L., and Stuhl, O.: Silylated Synthons. Facile Organic Reagents of Great Applicability. *88*, 33–88 (1980).

Boettcher, R. J., see Mislow, K.: *47*, 1–22 (1974).

Bolletta, F., see Balzani, V.: *75*, 1–64 (1978).

Brandmüller, J., and Schrötter, H. W.: Laser Raman Spectroscopy of the Solid State. *36*, 85–127 (1973).

Bremser, W.: X-Ray Photoelectron Spectroscopy. *36*, 1–37 (1973).

Breuer, H.-D., see Winnewisser, G.: *44*, 1–81 (1974).

Brewster, J. H.: On the Helicity of Variously Twisted Chains of Atoms. *47*, 29–71 (1974).

Brocas, J.: Some Formal Properties of the Kinetics of Pentacoordinate Stereoisomerizations. *32*, 43–61 (1972).

Brown, H. C.: Meerwein and Equilibrating Carbocations. *80*, 1–18 (1979).

Brunner, H.: Stereochemistry of the Reactions of Optically Active Organometallic Transition Metal Compounds. *56*, 67–90 (1975).

Buchs, A., see Delfino, A. B.: *39*, 109–137 (1973).

Bürger, H., and Eujen, R.: Low-Valent Silicon. *50*, 1–41 (1974).

Burgermeister, W., and Winkler-Oswatitsch, R.: Complexformation of Monovalent Cations with Biofunctional Ligands. *69*, 91–196 (1977).

Burns, J. M., see Koch, T. H.: *75*, 65–95 (1978).

Butler, R. S., and deMaine, A. D.: CRAMS – An Automatic Chemical Reaction Analysis and Modeling System. *58*, 39–72 (1975).

Caesar, F.: Computer-Gas Chromatography. *39*, 139–167 (1973).

Capitelli, M., and Molinari, E.: Kinetics of Dissociation Processes in Plasmas in the Low and Intermediate Pressure Range. *90*, 59–109 (1980).

Carreira, A., Lord, R. C., and Malloy, T. B., Jr.: Low-Frequency Vibrations in Small Ring Molecules *82*, 1–95 (1979).

Čársky, P., and Zahradník, R.: MO Approach to Electronic Spectra of Radicals. *43*, 1–55 (1973).

Čársky, P., see Hubač, J.: *75*, 97–164 (1978).

Caubère, P.: Complex Bases and Complex Reducing Agents. New Tools in Organic Synthesis. *73*, 49–124 (1978).

Chan, K., see Venugopalan, M.: *90*, 1–57 (1980).

Chandra, P.: Molecular Approaches for Designing Antiviral and Antitumor Compounds. *52*, 99–139 (1974).

Chandra, P., and Wright, G. J.: Tilorone Hydrochloride. The Drug Profile. *72*, 125–148 (1977).

Chapuisat, X., and Jean, Y.: Theoretical Chemical Dynamics: A Tool in Organic Chemistry. *68*, 1–57 (1976).

Cherry, W. R., see Epiotis, N. D.: *70*, 1–242 (1977).

Chini, P., and Heaton, B. T.: Tetranuclear Clusters. *71*, 1–70 (1977).

Christian, G. D.: Atomic Absorption Spectroscopy for the Determination of Elements in Medical Biological Samples. *26*, 77–112 (1972).

Clark, G. C., see Wasserman, H. H.: *47*, 73–156 (1974).

Clerc, T., and Erni, F.: Identification of Organic Compounds by Computer-Aided Interpretation of Spectra. *39*, 91–107 (1973).

Clever, H.: Der Analysenautomat DSA-560. *29*, 29–43 (1972).

Connor, J. A.: Thermochemical Studies of Organo-Transition Metal Carbonyls and Related Compounds. *71*, 71–110 (1977).

Connors, T. A.: Alkylating Agents. *52*, 141–171 (1974).

Craig, D. P., and Mellor, D. P.: Discriminating Interactions Between Chiral Molecules. *63*, 1–48 (1976).

Cram, D. J., and Cram, J. M.: Stereochemical Reaction Cycles. *31*, 1–43 (1972).

Cresp, T. M., see Sargent, M. V.: *57*, 111–143 (1975).

Crockett, G. C., see Koch, T. H.: *75*, 65–95 (1978).

Dauben, W. G., Lodder, G., and Ipaktschi, J.: Photochemistry of β,γ-unsaturated Ketones. *54*, 73–114 (1974).

DeClercq, E.: Synthetic Interferon Inducers. *52*, 173–198 (1974).

Degens, E. T.: Molecular Mechanisms on Carbonate, Phosphate, and Silica Deposition in the Living Cell. *64*, 1–112 (1976).

Delfino, A. B., and Buchs, A.: Mass Spectra and Computers. *39*, 109–137 (1973).

DeLuca, H. F., Paaren, H. E., and Schnoes, H. K.: Vitamin D and Calcium Metabolism. *83*, 1–65 (1979).

DeMaine, A. D., see Butler, R. S.: *58*, 39–72 (1975).

DePuy, C. H.: Stereochemistry and Reactivity in Cyclopropane Ring-Cleavage by Electrophiles. *40*, 73–101 (1973).

Inorganic and Physical Chemistry

1978. 158 figures, 25 tables. IV, 239 pages.
(Topics in Current Chemistry, Volume 77)
ISBN 3-540-08987-X

Contents:
J. J. Bikerman: Surface Energy of Solids.
(187 ref.)
H. G. Wiedemann, G. Bayer: Trends and Applications of Thermogravimetry. (69 ref.)
M. B. Huglin: Determination of Molecular
Weights by Light Scattering. (176 ref.)

Structure of Liquids

1975. 88 figures, 38 tables. IV, 205 pages.
(Topics in Current Chemistry, Volume 60)
ISBN 3-540-07484-8

Contents:
P. Schuster, W. Jakubetz, W. Marius: Molecular Models for the Solvation of Small Ions
and Polar Molecules. (268 ref.)
S. A. Rice: Conjectures on the Structure of
Amorphous Solid and Liquid Water. (94 ref.)

Bonding and Structure

1976. 55 figures, 25 tables. IV, 202 pages.
(Topics in Current Chemistry, Volume 63)
ISBN 3-540-07605-0

Contents:
D. P. Craig, D. P. Mellor: Discriminating Interactions Between Chiral Molecules. (64 ref.)
R. Gleiter, R. Gygax: No-Bond-Resonance
Compounds, Structure, Bonding and Properties. (147 ref.)
D. H. Sutter, W. H. Flygare: The Molecular
Zeeman Effect. (77 ref.)

Springer-Verlag
Berlin
Heidelberg
New York

K. L. Kompa

Chemical Lasers

1973. 31 figures. III, 92 pages.
(Fortschritte der chem. Forschung, Band 37)
ISBN 3-540-06099-5

Contents:
Population Inversion and Molecular Amplification. Energy-Partitioning in Elementary
Chemical Reactions. Vibrational Relaxation.
Requirements for Laser Oscillation. Design
Parameters of Pulsed Chemical Lasers.
Specific Chemical Laser Systems. Future
Chemical Lasers. Present Perspectives of
High-Power Chemical Lasers. Kinetic Information through Chemical Laser Studies.
(205 ref.)

Inorganic Chemistry

E. Fluck , J. R. Wasson, G. M. Woltermann,
H. J. Stoklosa

1973. 15 figures. III, 129 pages.
(Fortschritte der chem. Forschung, Band 35)
ISBN 3-540-06080-4

Contents:
E. Fluck: The Chemistry of Phosphine.
(492 ref.)
J. R. Wasson, G. M. Woltermann, H. J. Stoklosa:
Transition Metal Dithio- and Diselenophosphate Complexes. (428 ref.)

W. Demtröder

Laser Spectroscopy

2nd, enlarged edition. 1973. 16 figures,
3 tables. III, 106 pages.
(Fortschritte der chem. Forschung, Band 17)
ISBN 3-540-06334-X

Contents:
Spectroscopy with Lasers: Introduction.
Characteristic Features of Lasers as Spectroscopic Light Sources. Spectroscopic Applications of Lasers. High-Resolution Spectroscopy Based on Saturation Effects. Spectroscopy of Laser Media. Conclusion. Zusammenfassung. (418 ref.)

A. F. Williams

A Theoretical Approach to Inorganic Chemistry

1979. 144 figures, 17 tables.
XIII, 316 pages
ISBN 3-540-09073-8

Contents:

This book outlines the application of simple quantum mechanics to the study of inorganic chemistry, and shows its potential for systematizing and understanding the structure, physical properties, and reactivities of inorganic compounds. The considerable strides made in inorganic chemistry in recent years necessitate the establishment of a theoretical framework if the student is to acquire a sound knowledge of the subject. A wide range of topics is covered, and the reader is encouraged to look for further extensions of the theories discussed. The book emphasizes the importance of the critical application of theory and, although it is chiefly concerned with molecular orbital theory, other approaches are discussed. This text is intended for students in the latter half of their undergraduate studies. (235 references)

From the Foreword
by Prof. C. K. Jørgensen

"... Dr. Alan Williams has acquired a considerable experience in work with transition metal complexes at the Universities of Cambridge and Geneva. In this book he has tried to avoid the variety of ephemeral and often contradictory rationalisations encountered in this field, and has made a careful comparison of modern opinions about chemical bonding. In my opinion this effort is fruitful for all students and active scientists in the field of inorganic chemistry. The distant relations to group theory, atomic spectroscopy and epistemology are brought into daylight.

... The interdisciplinary approach of the book shows up in the careful consideration given to many experimental techniques such as vibrational (infra-red and Raman), electronic (visible and ultraviolet), Mössbauer, magnetic resonance, and photoelectron spectra, with data for gaseous and solid samples as well as selected facts about solution chemistry. The book could not have been written a few years ago, and is likely to remain a highly informative survey of modern inorganic chemistry and chemical physics."

Springer-Verlag
Berlin
Heidelberg
New York